Advances in Asian Human-Environmental Research

The *Advances in Asian Human-Environmental Research* series aims at fostering the discussion on the complex relationships between physical landscapes, natural resources, and their modification by human land use in various environments of Asia. It is widely acknowledged that human-environmental interactions become increasingly important in Area Studies and development research, taking into account regional differences as well as bio-physical, socio-economic and cultural particularities. The book series seeks to explore theoretic and conceptual reflection on dynamic human-environment systems applying advanced methodology and innovative research perspectives. The main themes of the series cover urban and rural landscapes in Asia. Examples include topics such as land and forest degradation, vulnerability and mitigation strategies, natural hazards and risk management concepts, environmental change, impact studies and consequences for local communities. The relevant themes of the series are mainly focused on geographical research perspectives of Area Studies, however there is scope for interdisciplinary contributions from various spheres within the natural sciences, social sciences and humanities. Key themes: Human-Environment Interaction - Asian regional studies - Asian geography - Impact studies - Landscape - Society - Land use and land cover change - Natural resources *Submit a proposal*: Proposals for the series will be considered by the Series Editor and International Editorial Board. An initial author/editor questionnaire and instructions for authors can be obtained from the Publisher, Dr. Robert K. Doe (robert.doe@springer.com).

Pravat Kumar Shit · Md Hasanuzzaman

Forest Landscape Dynamics in West Bengal

A Social-Ecological Perspective

Springer

Pravat Kumar Shit
Department of Geography
Raja Narendralal Khan Women's College (Autonomous)
Midnapore, West Bengal, India

Md Hasanuzzaman
Department of Geography
Raja Narendralal Khan Women's College (Autonomous)
Midnapore, West Bengal, India

ISSN 1879-7180 ISSN 1879-7199 (electronic)
Advances in Asian Human-Environmental Research
ISBN 978-3-032-23699-9 ISBN 978-3-032-23700-2 (eBook)
https://doi.org/10.1007/978-3-032-23700-2

This Springer imprint is published by the registered company Springer Nature Switzerland AG
The registered company address is: Gewerbestrasse 11, 6330 Cham, Switzerland

Dedicated to beloved teachers and parents

Foreword

The past half-century represents a profound rupture in the relationship between humans and the natural world. The doubling of the global population alongside a fourfold expansion of the world economy reflects not merely demographic and economic trends, but the dominance of an anthropocentric worldview one that places human needs, desires, and progress at the centre of all value systems. Within this paradigm, nature is reduced to a reservoir of resources, instrumental to human advancement rather than possessing intrinsic worth.

Such unprecedented growth has dramatically intensified the demand for food, energy, and raw materials, accelerating processes of extraction, production, and consumption. Environmental philosophers, particularly those aligned with deep ecology, argue that this trajectory reveals a fundamental ethical crisis: the failure to recognize the inherent value of non-human life and ecological systems. The biosphere is no longer seen as a community of interdependent beings but as a commodified landscape subject to human control.

Consequently, natural systems, climate regimes, forests, oceans, and biodiversity networks have been transformed to an extent that challenges their resilience and regenerative capacities. Many of these alterations are not only extensive but irreversible, raising serious concerns about intergenerational justice and the moral responsibility humans bear toward future life forms. The current ecological crisis thus underscores the limitations of growth-centric development models and calls for a paradigm shift toward ecocentrism, sustainability, and ethical restraint.

In this light, environmental philosophy urges a rethinking of progress itself. Rather than equating well-being with material expansion, it advocates for a more harmonious coexistence with nature, grounded in respect, balance, and the recognition of ecological limits.

It is with deep admiration that I write this foreword to *Forest Landscape Dynamics in West Bengal: A Social-Ecological Perspective*. At a critical moment in history, when tropical forests face intense ecological and political pressures, this volume offers timely, evidence-based insights. It dexterously examines the interplay between ecological processes and human interventions, presenting forests as dynamic social-ecological systems. By linking regional realities with global environmental challenges, the book advances debates on sustainability, governance, and justice. It is a valuable contribution for scholars, policymakers, and practitioners committed to fostering balanced and responsible human-nature relationships.

This book represents a significant scientific milestone in that direction. It offers a comprehensive and integrative understanding of why forests matter, what challenges they face, and how forest-based solutions can contribute to building a climate-neutral and socially resilient ecological system. What distinguishes this work is its rare intellectual integrity: the authors refuse to draw a boundary between scientific analysis and human realities. Instead, they skill-fully weave both dimensions into a coherent and compelling narrative.

Grounded in methodological rigor, the study employs advanced tools such as remote sensing, Geographic Information Systems (GIS), machine learning, and extensive field-based surveys. Through these approaches, the authors meticulously document the troubling trajectory of forest landscapes in West Bengal, including a 30.2% decline in forest cover, a 49% reduction in ecosystem service values, and the progressive fragmentation of critical ecological corridors, particularly around the Siliguri region. These figures are not mere statistics; they signify the gradual erosion of ecological systems that sustain more than three million people and play a vital role in sequestering approximately 87.5 million tonnes of carbon.

Yet, the strength of this book lies not only in its analytical depth but also in its human sensitivity. The authors move beyond quantitative assessments to uncover the historical, social, and political dimensions of forest transformation. From the colonial legacy embedded in the Forest Acts of 1865 and 1878, to the complex livelihood realities of tribal communities in Binpur-II, and the persistent gender disparities within Joint Forest Management institutions-where women's participation in grievance redressal mechanisms remains critically low-this work underscores the inseparability of ecological sustainability and social justice.

Case studies such as the *Arabari* experiment, the gradual submergence of the Sundarbans, and biodiversity loss in Bankura collectively illustrate the broader consequences of governance failures and policy disconnects. These narratives powerfully demonstrate that forests are not merely ecological entities but lived landscapes, deeply intertwined with culture, identity, and survival.

I warmly commend Pravat Kumar Shit and Md Hasanuzzaman for producing a work of enduring academic and practical significance. This book will serve as an indispensable resource for researchers, planners, policymakers, and students across disciplines such as geography, environmental science, forestry, and development studies. More importantly, it offers not just a diagnosis of crisis, but a forward-looking framework-a compass guiding us toward integrated, equitable, and climate-adaptive forest governance in West Bengal and across the Global South.

April 2026

Prof. Nilanjana Das Chatterjee
Department of Geography
Vidyasagar University
Midnapore, India

Preface

Forest ecosystems act as vital carbon sinks and play a crucial role in reducing greenhouse gas concentrations. Beyond their environmental importance, forests significantly contribute to the socio-economic well-being of developing countries. Although the livelihoods of people living on forest fringes have long been dominated by subsistence agriculture, non-timber forest products (NTFPs) continue to play a critical role in meeting the basic needs of tribal communities.

Against this background, this book highlights spatial modeling approaches in forest resource issues, societal development, and forest landscape restoration. It focuses on sustainable management practices and presents research observations related to forest health, forest biomass, carbon stocks, and climate change. Forest ecosystems face numerous challenges and uncertainties arising from the combined effects of natural variability, climate change, and human-induced pressures. These complexities make decision-making more intricate and directly influence the sustainable utilization and allocation of current and future forest resources.

Consequently, it is essential to integrate advanced technologies and analytical tools into forest management and planning. In this regard, geospatial datasets, advanced image processing techniques, GIS-based analytical platforms, and programming languages such as R can significantly contribute to effective mapping, monitoring, and interpretation of land surface processes. Land sustainability strategies, forest biomass assessment, carbon stock estimation, soil-water resource management, and climate change mitigation along with informed policy development must be grounded in accurate surveys and assessments. Techniques based on remote sensing (RS), GIS, and R programming will play an increasingly important role in shaping appropriate management policies and plans.

This volume advances scientific understanding of geospatial technologies and their application in forest resource mapping, monitoring, and modeling, using up-to-date remote sensing and GIS methodologies for sustainable resource management. It provides a critical analysis of the debates and discourses surrounding forest resources and society; explores the intricate relationships between forests, forest fringe communities, and their livelihoods; and presents novel, region-specific strategies for sustainable management supported by detailed case studies.

The book is organized into ten chapters. Each chapter begins with a brief introduction outlining core concepts, followed by focused discussions on thematic research, emerging knowledge, and relevant case studies. This book will be of interest to forest resource planners and decision-makers, academic researchers, policymakers, non-governmental organizations, as well as undergraduate and postgraduate students in Geography, Botany, Wildlife Management, Remote Sensing & GIS, Forestry, and Environmental Sciences.

Midnapore, West Bengal, India

Pravat Kumar Shit
Md Hasanuzzaman

Acknowledgements

We are deeply obliged to the forest dwellers for their kind support and the valuable time they devoted during the field survey and household survey. We are sincerely thankful to the forest officers and staff for sharing their experiences, offering useful suggestions, and providing continuous encouragement and immense support throughout the course of this work.

Our heartfelt thanks go to Dr. Doris Bleier, whose encouragement, affection, and unwavering support kept us motivated until the book reached its final form. This work has been several years in the making, and we are therefore grateful to our families and friends for their continuous support and patience.

Dr. Pravat Kumar Shit expresses his sincere thanks to Dr. Swapna Ghorai, Principal, Raja N. L. Khan Women's College (Autonomous), Midnapore, for her administrative support in carrying out this project. We also acknowledge the Department of Geography, Raja N. L. Khan Women's College (Autonomous) for providing logistical support and essential infrastructure facilities.

This work would not have been possible without the constant inspiration from our students, the invaluable lessons from our teachers, the enthusiasm of our colleagues and collaborators, and the unwavering support of our families.

Midnapore, West Bengal, India

Pravat Kumar Shit
Md Hasanuzzaman

Contents

Abbreviations

AGB	Above-Ground Biomass
ALS	Airborne Laser Scanning
ARIES	Artificial Intelligence for Ecosystem Services
AUC	Area Under Curve
BOA	Bottom of Atmosphere
CBA	Community-Based Adaptation
CD Block	Community Development (CD) Block
CFR	Community Forest Resource
CLEAR	Center for Land Use Education and Research
DBH	Diameter at Breast Height
ESA	European Space Agency
ESP	Ecosystem Service Potentiality
ESVs	Ecosystem Service Values
EVI	Enhanced Vegetation Index
FAO	Food and Agriculture Organization
FCAM	Forest Carbon Assessment and Monitoring
FGDs	Focus Group Discussions
FPCs	Forest Protection Committees
FRA	Forest Rights Act
GCPs	Ground Control Points
GEDI	Global Ecosystem Dynamics Investigation
GEE	Google Earth Engine
GIS	Geographic Information Systems
GLCM	Gray Level Co-occurrence Matrix
GPS	Global Positioning System
HFHRA	High Forest and High Road Access
HFLRA	High Forest and Low Road Access
ICDS	Integrated Child Development Services
ICZM	Integrated Coastal Zone Management
IK	Indigenous Knowledge Systems
ILWIS	Integrated Land and Water Information System

IMD	India Meteorological Department
InVEST	Integrated Valuation of Ecosystem Services and Trade-offs
ISFR	India State of Forest Report
JFM	Joint Forest Management
KMA	Kolkata Metropolitan Area
LAI	Leaf Area Index
LCM	Land Change Modeler
LDI	Livelihood Dependency Index
LFHRA	Low Forest and High Road Access
LFLRA	Low Forest and Low Road Access
LFT	Landscape Fragmentation Tool
LiDAR	Light Detection and Ranging
LULC	Land Use Land Cover
MAE	Mean Absolute Error
MCE	Multi-Criteria Evaluation
MODIS	Moderate Resolution Imaging Spectroradiometer
MLC	Maximum Likelihood Classification
MoEF & CC	Ministry of Environment, Forest and Climate Change
NBR	Normalized Burn Ratio
NDVI	Normalized Difference Vegetation Index
NFHS	National Family Health Survey
NIR	Near-Infrared
NTFPs	Non-Timber Forest Products
OLI	Operational Land Imager
QGIS	Quantum Geographic Information System
REDD+	Reducing Emissions from Deforestation and Forest Degradation
RF	Random Forest
RMSE	Root Mean Square Error
ROC	Receiver Operating Characteristic
SAR	Synthetic Aperture Radar
SCs	Scheduled Castes
SDGs	Sustainable Development Goals
SES	Social-Ecological Systems
SFD	State Forest Departments
SHGs	Self-Help Groups
SIMUS	System for Integrated Modeling of Urban Systems
SLI	Standard of Living Index
STs	Scheduled Tribes
SWIR	Short-Wave Infrared
TEK	Traditional Ecological Knowledge
TM	Thematic Mapper
UNESCO	United Nations Educational, Scientific and Cultural Organization
USGS	United States Geological Survey
VGI	Volunteered Geographic Information

List of Figures

List of Tables

Chapter 1
Introduction

Abstract Accelerating deforestation, habitat fragmentation, and socio-political transformations have rendered tropical forest landscapes among the most contested social-ecological systems globally. India's West Bengal, encompassing approximately 16,902 km^2 of ecologically diverse forests spanning Sundarbans mangroves, sal-dominated Jangal Mahal woodlands, and Himalayan foothill forests, exemplifies compounding pressures where colonial governance legacies, post-independence developmental imperatives, metropolitan expansion, and climate variability converge to threaten both biodiversity and forest-dependent tribal livelihoods. This study comprehensively investigates the spatiotemporal dynamics of West Bengal's forest landscapes through an integrated social-ecological systems (SES) lens, examining historical governance transitions, fragmentation patterns, carbon stock distribution, ecosystem service provision, and participatory forest management outcomes. A rigorous multi-method framework was employed, integrating multi-temporal Landsat and Sentinel-2 satellite imagery, GIS-based fragmentation metrics, machine learning classification algorithms, InVEST-based ecosystem service valuation, allometric biomass estimation, stratified household surveys, participatory rural appraisal, and institutional policy analysis across ecologically and socially diverse districts. Findings reveal dramatic forest cover decline from approximately 30 to 19.07%, intensifying fragmentation particularly within peri-urban corridors, significant carbon stock heterogeneity across forest types, escalating urban-forest conflicts near Siliguri, and mixed equity outcomes within Joint Forest Management (JFM) schemes including persistent elite capture and gender exclusion despite measurable ecological recovery. Achieving sustainable forest landscapes necessitates genuinely decentralized polycentric governance that authentically integrates traditional ecological knowledge, strengthens community tenure rights under the Forest Rights Act 2006, and adopts climate-adaptive landscape planning. These findings offer transferable frameworks for forest conservation policy across the Global South.

Keywords Forest fragmentation · Remote sensing · Social-ecological systems · Traditional ecological knowledge: Joint Forest Management · Carbon stocks · Ecosystem services · Land-use change · Participatory governance · West Bengal

P. K. Shit and M. Hasanuzzaman, *Forest Landscape Dynamics in West Bengal*,
Advances in Asian Human-Environmental Research,
https://doi.org/10.1007/978-3-032-23700-2_1

1.1 Global Context: Forests at the Crossroads

Forests constitute one of Earth's most critical biomes, covering approximately 4.06 billion hectares or roughly 31% of the global land surface (FAO 2020). These ecosystems harbor an estimated 80% of terrestrial biodiversity, store nearly 861 gigatons of carbon, and provide essential services to billions of people worldwide (Pan et al. 2011; Barlow et al. 2016). Yet forests stand at a critical juncture in human history. The Global Forest Resources Assessment reveals that approximately 10 million hectares of forest are lost annually, with tropical regions bearing the brunt of this degradation (FAO 2020). This persistent deforestation, coupled with forest degradation through fragmentation, selective logging, and climate change impacts, poses profound threats to planetary ecological stability and human well-being. The relationship between human societies and forest ecosystems has evolved dramatically over millennia, from hunter-gatherer dependencies through agricultural expansion to contemporary industrial exploitation and conservation initiatives (Mather 2007; Ritter and Dauksta 2013). Traditional societies often maintained sustainable relationships with forests, developing sophisticated management systems rooted in traditional ecological knowledge (Agnoletti 2007). However, the acceleration of deforestation during the colonial era and subsequent post-independence industrial development fundamentally altered forest landscapes across much of the developing world (Roy and Fleischman 2022).

Contemporary forest challenges extend beyond simple deforestation to encompass complex interactions between land-use change, climate variability, biodiversity loss, and socio-economic transformations (Lambin and Meyfroidt 2011; Curtis et al. 2018). Forest fragmentation-the subdivision of continuous forest cover into smaller, isolated patches has emerged as a critical threat, with over 70% of remaining forests now lying within one kilometer of a forest edge (Haddad et al. 2015). Such fragmentation profoundly affects ecosystem functions, disrupting wildlife corridors, altering microclimates, reducing carbon storage capacity, and diminishing the provision of ecosystem services upon which rural populations depend (Fischer and Lindenmayer 2007; Fahrig 2017).

Simultaneously, rapid urbanization creates intensifying conflicts between expanding metropolitan areas and adjacent forest ecosystems. With global urban populations projected to reach 68% by 2050, urban sprawl increasingly encroaches upon forests that provide critical ecosystem services including climate regulation, water purification, and carbon sequestration (Arora and Mishra 2019). This urban-forest interface generates complex management challenges requiring integrated landscape approaches that reconcile development imperatives with ecological conservation (Seto et al. 2012; Grimm et al. 2008).

Climate change adds another layer of complexity to forest governance. Rising temperatures, altered precipitation patterns, and increased frequency of extreme events interact synergistically with anthropogenic pressures to amplify forest vulnerability (Allen et al. 2010; Anderegg et al. 2020). Simultaneously, forests represent crucial natural climate solutions, with forest conservation and restoration potentially contributing up to 30% of the mitigation needed to limit global warming to 2 °C (Griscom et al. 2017). Understanding and managing these coupled social-ecological dynamics has become imperative for achieving both conservation objectives and sustainable development goals.

1.2 Forests of India: Diversity, Degradation, and Governance

India's forests, spanning approximately 71.22 million hectares or 21.71% of the nation's geographical area (Fig. 1.1), represent extraordinary ecological diversity (Forest Survey of India 2021). From the tropical rainforests of the Western Ghats and Northeast India through the dry deciduous forests of central India to the temperate and alpine forests of the Himalayas, these ecosystems support remarkable biodiversity including over 45,000 plant species and diverse fauna (Rawal et al. 2013). The forests provide livelihood support to approximately 275 million people, many belonging to marginalized tribal and forest-dwelling communities whose survival remains intimately connected to forest resources (MoEFCC, 2019).

However, India's forest history reflects profound transformations shaped by colonialism, post-independence development policies, and contemporary conservation initiatives (Roy and Fleischman 2022). The colonial period witnessed systematic exploitation of forest resources to serve imperial economic interests, fundamentally restructuring forest governance through legislation like the Indian Forest Act of 1878 that converted community-managed resources into state property (Gautam 2020). This transition criminalized traditional forest use, marginalized indigenous management systems, and established centralized bureaucratic control that persisted well into the post-independence era.

Post-independence forest policy initially prioritized industrial supply and revenue generation, continuing colonial extractive patterns (Domínguez and Luoma 2020). The early decades saw accelerated deforestation driven by agricultural expansion, infrastructure development, and industrial demands. India's forest cover declined from an estimated 40-50% in the pre-colonial era to approximately 19% by the 1980s, representing a massive ecological transformation (Richards and Flint 1994; Roy and Fleischman 2022). This degradation prompted growing environmental consciousness and social movements advocating for participatory forest governance.

The 1970s and 1980s marked a paradigm shift in Indian forest policy. Grassroots movements like Chipko and scientific critiques of centralized forestry challenged the prevailing exploitation model (Guha 2000). The National Forest Policy of 1988 represented a conceptual breakthrough, prioritizing ecological stability and community participation over revenue generation. Subsequently, Joint Forest Management (JFM) emerged as India's primary participatory forest governance model, pioneered through the Arabari experiment in West Bengal (Poffenberger and McGean 1996; Dey and Arunachalam 2024).

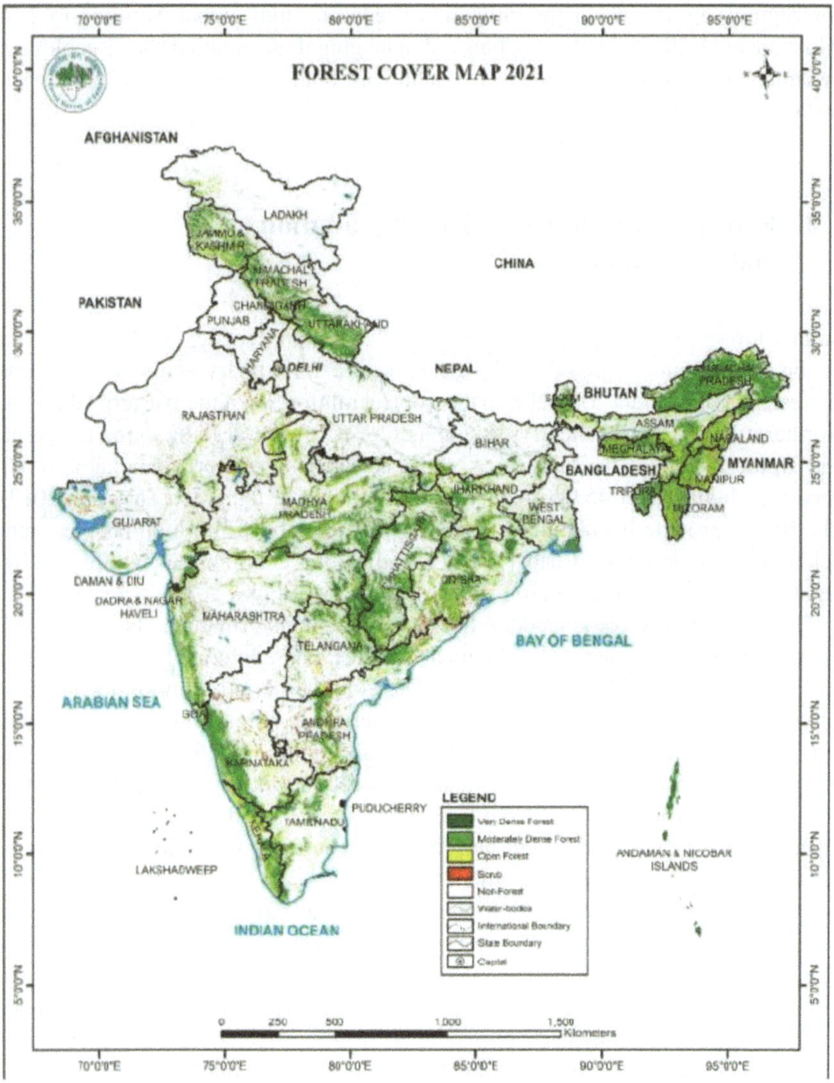

Fig. 1.1 Forest cover map of India. *Source* India state of forest report (ISFR) 2021

More recently, the Scheduled Tribes and Other Traditional Forest Dwellers (Recognition of Forest Rights) Act of 2006 commonly known as the Forest Rights Act sought to legally recognize individual and community forest rights, acknowledging historical injustices and empowering forest-dependent communities (Kumar and Kerr 2012; Springate-Baginski and Blaikie 2007). However, implementation has remained uneven across states, with bureaucratic resistance, elite capture, and inadequate institutional support constraining the realization of these legal provisions (Kashwan 2017).

Contemporary Indian forests face multiple pressures including mining, infrastructure expansion, urban sprawl, and climate change (Khosla and Bhardwaj 2019). Understanding these dynamics requires moving beyond simple forest cover statistics to examine spatial patterns of fragmentation, changes in ecosystem service provision, carbon dynamics, and livelihood dependencies (Oldekop et al. 2020; Ramachandra et al. 2020). Geographic Information Systems (GIS) and remote sensing technologies have revolutionized such assessments, enabling precise monitoring of land-use change, biomass estimation, and biodiversity mapping at multiple spatial and temporal scales (Nizamani et al. 2024; Pandey et al. 2022).

1.3 West Bengal's Forest Landscapes: A Small-Scale Example of Social-Ecological Complexity

West Bengal, situated in eastern India, presents a compelling case study for examining coupled social-ecological dynamics in forest landscapes (Sen 2021) (Fig. 1.2). The state's forests, spanning approximately 16,902 km^2 or roughly 19% of its geographical area, encompass remarkable ecological diversity ranging from the world's largest mangrove ecosystem in the Sundarbans through tropical moist and dry deciduous forests in the western plains to subtropical and temperate forests in the northern hills (Forest Survey of India 2021). This diversity reflects West Bengal's position at the confluence of multiple biogeographic zones and climatic regimes.

The Sundarbans mangroves, a UNESCO World Heritage Site covering approximately 4200 km^2, constitute one of the planet's most critical ecosystems, providing coastal protection, supporting fisheries, harboring the iconic Royal Bengal tiger, and storing exceptional carbon stocks estimated at 400–600 Mg C per hectare (Alongi 2008; Giri et al. 2011). However, these mangroves face mounting pressures from aquaculture expansion, agricultural encroachment, sea-level rise, and intensifying cyclones dynamics that exemplify the climate vulnerability of coastal forest ecosystems (Ghosh et al. 2015; Mukhopadhyay and Khan 2015).

The western districts of Bankura, Purulia, Paschim Medinipur, and Jhargram collectively known as Jangal Mahal contain extensive sal (*Shorea robusta*)-dominated tropical dry deciduous forests (Das 2021). These forests, constituting approximately 35% of the state's forest cover, support rich biodiversity and provide critical livelihood resources for predominantly tribal populations including the Santhal, Munda, and Lodha communities (Pal et al. 2024). Non-timber forest products such as sal leaves and seeds, mahua (*Madhuca indica*) flowers, kendu leaves, honey, and medicinal plants contribute substantially to household incomes, with forest resources accounting for 25–40% of tribal family earnings (Kumar and Saikia 2020).

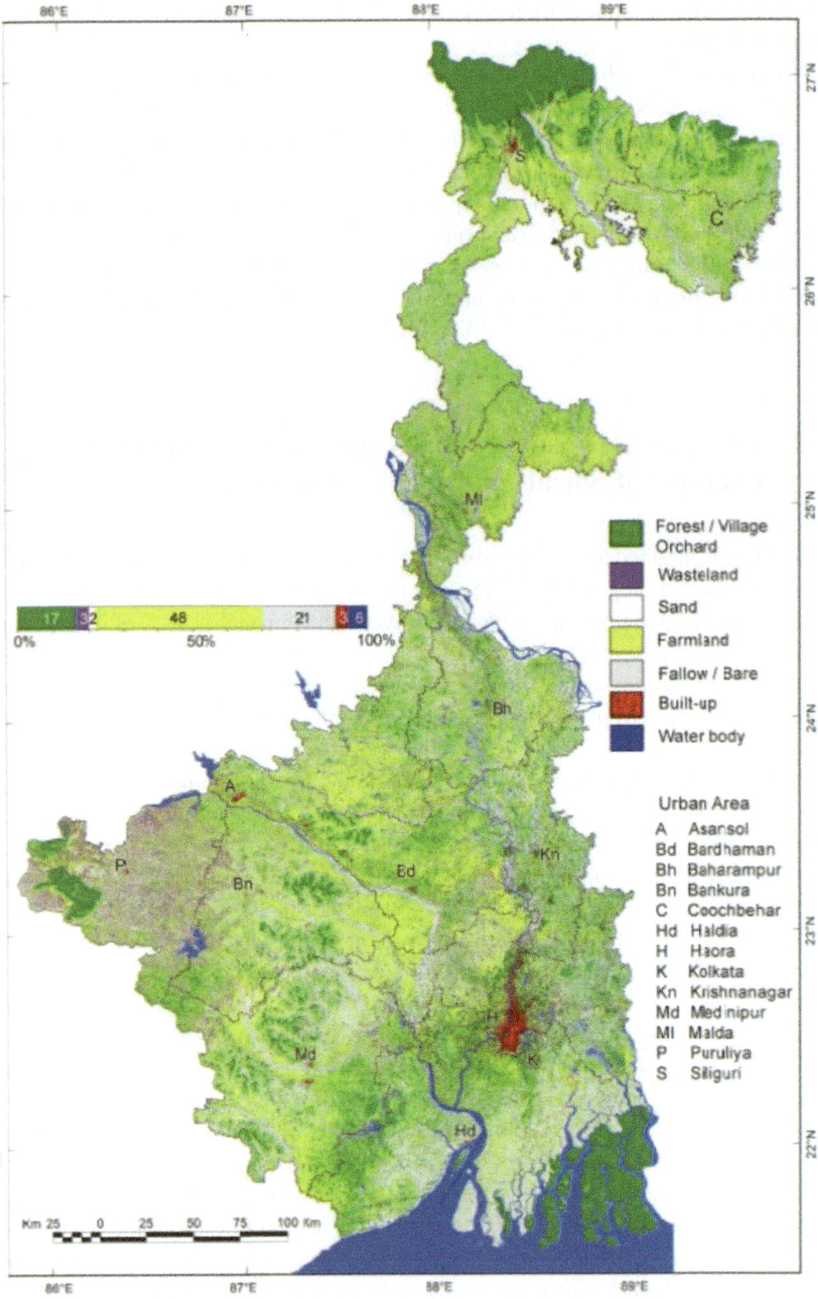

Fig. 1.2 Land-use and land-cover map of West Bengal. *Source* Modified after Bandyopadhyay et al. (2014)

The northern districts of Darjeeling, Kalimpong, and Jalpaiguri contain subtropical and temperate forests that harbor endemic species including numerous orchids, rhododendrons, and medicinal plants, alongside fauna such as the red panda, clouded leopard, and diverse avifauna (Karthigeyan et al. 2022). The Terai-Duar region's forests form critical elephant corridors connecting protected areas across India, Nepal, and Bhutan, though these corridors face severe fragmentation from tea plantations, settlements, and infrastructure development (Kushwaha and Hazarika 2004).

West Bengal's Forest history epitomizes broader patterns of colonial exploitation and post-independence governance transitions. The colonial period witnessed systematic exploitation of sal forests for railway sleepers and shipbuilding, establishment of reserved forests that criminalized traditional use, and introduction of European silvicultural methods that prioritized timber yield over ecological integrity (Sivaramakrishnan 1999). Resistance movements including the Santhal Rebellion (1855–56) and Chuar Uprising reflected tribal communities' opposition to dispossession, though these were ultimately suppressed through colonial military power (Kamra 2016).

Post-independence forest policy initially continued centralized control through retention of the Indian Forest Act of 1927 and further consolidation via the West Bengal Estate Acquisition Act of 1953 (Poffenberger and McGean 1996). Forest cover declined precipitously from approximately 30% in the early twentieth century to 13.4% by 1988, driven by agricultural expansion (accounting for 67% of forest loss), refugee resettlement following Partition, Green Revolution intensification, urbanization, and industrial development (Roy and Tomar 2000).

However, West Bengal also pioneered India's participatory forest governance through the Arabari experiment initiated by forest officer A.K. Banerjee in 1971–72. This innovative collaboration between the Forest Department and 11 villages in Midnapore District demonstrated that severely degraded forests could regenerate through community protection when villagers received usufruct rights and benefit-sharing (Roy and Tomar 2000). The success of Arabari inspired statewide expansion of Joint Forest Management from 1989 to 90 onward, with thousands of Forest Protection Committees formed across degraded forest areas (Mitra 1995).

Despite these participatory initiatives, West Bengal's forests continue facing multiple pressures. Metropolitan expansion around Kolkata has driven approximately 35% forest loss since 1990, while industrial corridors between Kolkata and Durgapur have reduced forest cover in some areas to well below the national average (Dutta and Guchhait 2022; Sarkar et al. 2024). Mining activities in western districts, infrastructure expansion including the Asian Highway and railway projects, and hydroelectric developments in northern hills generate additional pressures (Mondal and Southworth 2010). Forest fragmentation has intensified dramatically. Such fragmentation produces cascading ecological consequences including reduced species diversity, disrupted wildlife movement, intensified human-wildlife conflict, and diminished ecosystem service provision. Large mammals including elephants and gaurs have lost their habitat range, while mammalian diversity in small patches has declined by nearly half (Gross et al. 2021; Ram et al. 2021).

1.4 Social-Ecological Systems Framework: Integrating Nature and Society

Understanding forest landscape dynamics requires moving beyond traditional disciplinary boundaries to embrace integrated social-ecological systems (SES) perspectives that recognize the inextricable connections between human societies and natural ecosystems (Berkes and Folke 1998; Ostrom 2009) (Fig. 1.3). Social-ecological systems are characterized by complex feedbacks, non-linear dynamics, threshold effects, and adaptive cycles that make them fundamentally different from systems that can be understood through simple linear cause-effect relationships (Holling 2001; Walker et al. 2004).

The social-ecological systems framework conceptualizes forests not merely as ecological entities but as coupled human-nature systems where ecological processes and social dynamics continuously interact and co-evolve (Folke et al. 2010). From this perspective, forest change cannot be understood solely through biophysical variables but must incorporate governance institutions, power relations, economic incentives, cultural values, and historical legacies that shape how societies interact with forest landscapes (Ostrom and Cox 2010).

Several core concepts from social-ecological systems theory prove particularly relevant for forest landscape analysis. Resilience the capacity of a system to absorb disturbance while maintaining essential functions and structures provides crucial insights into how forest ecosystems and dependent communities respond to pressures

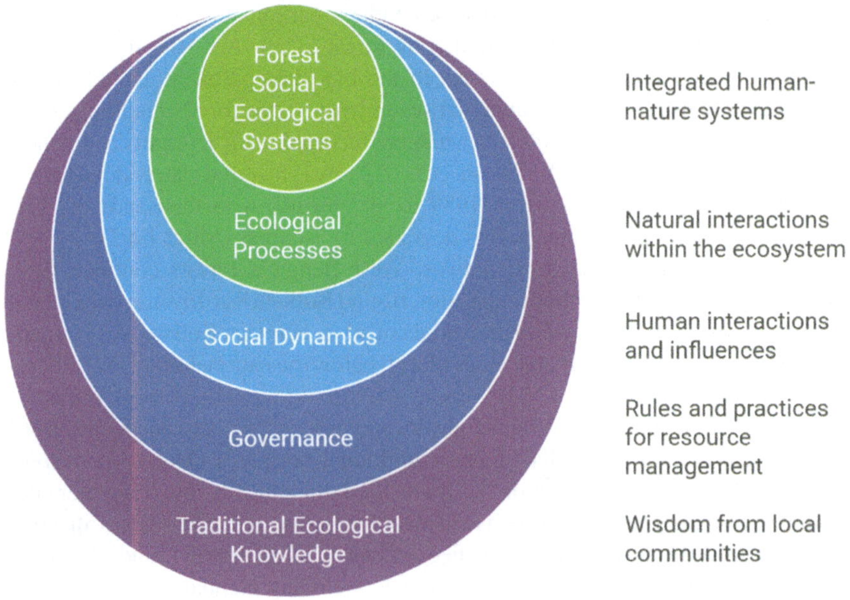

Fig. 1.3 Symmetrical diagram of forest socio-ecological systems

ranging from climate variability to policy changes (Walker et al. 2004). Adaptive capacity the ability of actors to influence resilience through learning, experimentation, and innovation helps explain differential responses to environmental change across communities and regions (Folke et al. 2003).

Ecosystem services the benefits people obtain from ecosystems provide a conceptual bridge connecting ecological processes to human well-being (Millennium Ecosystem Assessment 2005; Díaz et al. 2018). Forests provide provisioning services including timber, fuelwood, and non-timber forest products; regulating services such as climate moderation, water purification, and erosion control; cultural services encompassing recreation, aesthetics, and spiritual values; and supporting services including nutrient cycling and primary productivity (Costanza et al. 1997; Bisht et al. 2024). Understanding how forest change affects ecosystem service flows and how these changes impact different social groups is essential for designing equitable and sustainable management strategies.

Power relations and environmental justice considerations are central to social-ecological systems analysis. Forests are not neutral spaces but contested terrains where different actors state agencies, local communities, commercial interests, conservation organizations pursue competing visions of appropriate use and management (Peluso 1992; Sen 2021). Historical processes of dispossession, contemporary patterns of elite capture, and unequal distributions of costs and benefits from conservation and development initiatives all shape forest outcomes (Fairhead and Leach 1996; Agrawal and Gibson 1999).

Governance the formal and informal rules, norms, and practices that structure resource access and management emerges as critical for social-ecological sustainability (Dietz et al. 2003; Armitage et al. 2012). Effective governance requires balancing multiple objectives including ecological integrity, livelihood security, cultural values, and economic development, while navigating inherent trade-offs and managing conflicts (Young 2002). Polycentric governance systems that distribute authority across multiple nested scales often prove more adaptive and resilient than purely centralized or decentralized arrangements (Ostrom 2010).

Traditional ecological knowledge (TEK) accumulated wisdom, practices, and beliefs that indigenous and local communities have developed through generations of intimate interaction with their environments represents a crucial but often marginalized knowledge system (Nelson 2014). Effective forest governance requires creating genuine spaces for knowledge co-production that respect TEK alongside scientific expertise, rather than imposing top-down expert prescriptions that ignore local understanding (Martin et al. 2010).

1.5 Methodological Approaches: Integrating Multiple Tools and Perspectives

Assessing forest landscape dynamics within a social-ecological framework requires integrating diverse methodological approaches that span biophysical measurement, geospatial analysis, socio-economic investigation, and participatory research (Fig. 1.4). This volume employs a multi-method strategy combining remote sensing and GIS technologies with household surveys, focus group discussions, key informant interviews, and participatory mapping to generate comprehensive understanding.

Remote sensing technologies enable systematic monitoring of forest cover change, fragmentation patterns, and vegetation conditions across large spatial extents and extended time periods (Turner et al. 2003; Wulder et al. 2018). Landsat satellite imagery, with its 30-m spatial resolution and continuous archive extending back to 1972, provides the foundation for land-use and land-cover change analysis (Woodcock et al. 2008). Vegetation indices including the Normalized Difference Vegetation Index (NDVI) and Enhanced Vegetation Index (EVI) enable assessment of forest health and productivity (Huete et al. 2002). More recent sensors including Sentinel-2 with 10-m resolution enhance spatial detail, while synthetic aperture radar (SAR) data penetrate cloud cover that limits optical imagery in tropical regions (Gorelick et al. 2017). Advanced classification algorithms including random forest, support vector machines, and deep learning approaches achieve high accuracy in distinguishing land-cover types and detecting changes (Pelletier et al. 2016; Ma et al. 2019). Fragmentation metrics quantify spatial patterns including patch size, edge

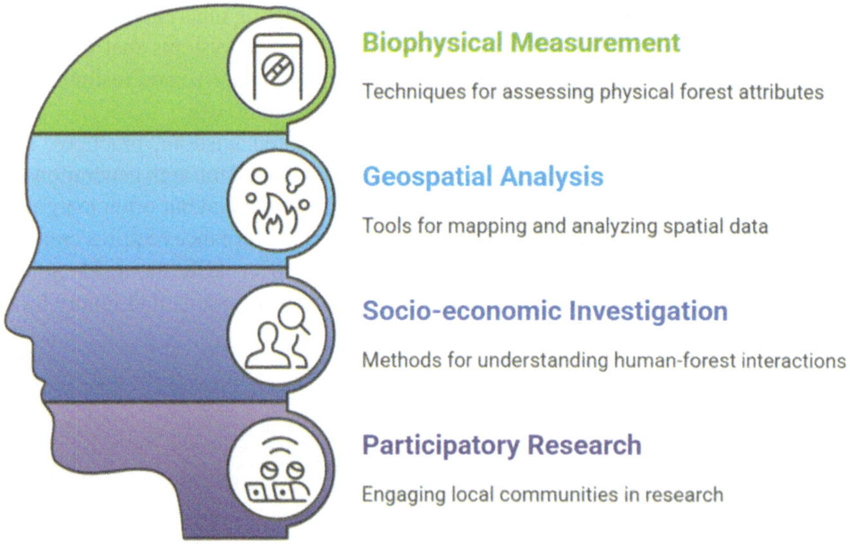

Fig. 1.4 Symmetrical diagram of a comprehensive forest landscape assessment

density, core area, and connectivity that influence ecological processes (Wang et al. 2014). Light Detection and Ranging (LiDAR) technology enables three-dimensional forest structure assessment crucial for biomass estimation (Xu et al. 2021).

Carbon stock estimation integrates multiple approaches including field measurements using allometric equations, remote sensing-derived vegetation indices, and machine learning models that synthesize diverse data sources (Rejou-Mechain et al. 2014; Baccini et al. 2017). Ecosystem service valuation combines biophysical modeling using tools like InVEST (Integrated Valuation of Ecosystem Services and Trade-offs) with economic valuation techniques to quantify service provision in monetary or physical units (Meraj et al. 2022). Such assessments reveal the substantial values forests provide often invisible in conventional economic accounting making explicit the trade-offs inherent in land-use decisions. Socio-economic research methods provide essential insights into human dimensions of forest landscapes. Household surveys employing stratified random sampling enable quantitative assessment of livelihood dependencies, income sources, forest resource utilization patterns, and perceptions of forest change (Lax and Köthke 2017). Standard of living indices, livelihood dependency metrics, and poverty assessments reveal how forest conditions affect human welfare across different social groups.

Qualitative methods including focus group discussions, semi-structured interviews, and participatory rural appraisal techniques capture nuanced understanding of community experiences, traditional knowledge, gender dynamics, and institutional functioning that surveys alone cannot reveal (Shoko et al. 2024). Participatory GIS approaches integrate local spatial knowledge with scientific mapping, empowering communities while generating insights into contested resources, customary tenure systems, and sacred spaces (Brown and Kyttä 2018). Policy and institutional analysis examines governance structures, legal frameworks, power relations, and implementation gaps that shape forest outcomes (Kimengsi et al. 2023). Document analysis, key informant interviews with officials and activists, and comparative institutional assessment illuminate how policies work in practice, often quite differently from their formal design. Validation and verification procedures ensure research reliability. Ground-truthing using GPS-referenced field observations verifies remote sensing classifications (Barrington et al. 2011). Cross-validation techniques assess model accuracy using independent datasets. Triangulation across multiple methods and data sources strengthens confidence in findings (Keevers 2019). Uncertainty quantification explicitly acknowledges limitations rather than presenting false precision.

1.6 Objectives and Structure of This Volume

This volume provides comprehensive analysis of West Bengal's Forest landscape dynamics from an integrated social-ecological perspective. Through systematic examination of historical transformations, contemporary ecological changes, livelihood dependencies, and governance innovations, it generates insights with relevance extending beyond West Bengal to broader questions of forest conservation and sustainable development in the Global South.

The specific objectives are as follows:

1. To trace the historical evolution of forest governance in West Bengal from pre-colonial indigenous systems through colonial exploitation to post-independence participatory initiatives, revealing how historical legacies shape contemporary conditions.
2. To quantify spatial and temporal patterns of forest fragmentation using advanced geospatial analysis, assessing ecological consequences including loss of core habitat, connectivity disruption, and edge effects.
3. To estimate forest biomass and carbon stocks across diverse forest types, evaluating West Bengal's role in climate change mitigation and identifying priority areas for conservation and restoration.
4. To analyze urban-forest conflicts resulting from metropolitan expansion, examining socio-ecological and economic dimensions while identifying strategies for integrated landscape planning.
5. To assess relationships between biodiversity, ecosystem services, and forest conditions, quantifying how fragmentation and degradation affect service provision and human welfare.
6. To investigate livelihood dependencies on forest resources, revealing spatial patterns, gender dimensions, and socio-economic constraints that shape forest-dependent communities' welfare.
7. To evaluate Joint Forest Management initiatives, examining both ecological recovery and social equity outcomes while identifying factors that facilitate or constrain participatory governance.
8. To analyze eco-development and sustainable forest management approaches, assessing their contributions to reconciling conservation with community welfare under conditions of climate change.

The volume is structured as follows. Chapter 2 examines historical transformations of forest cover from the colonial era to the present, tracing how different governance regimes shaped forest landscapes and community rights. Chapter 3 focuses on forest fragmentation, employing geospatial analysis to quantify patterns in Jhargram District and assess ecological consequences. Chapter 4 investigates forest biomass and carbon stocks, using field measurements and remote sensing to estimate storage across forest types with detailed analysis of the Jangal Mahal region. Chapter 5 analyzes urban expansion and forest conflicts, examining metropolitan encroachment with a case study of Siliguri Town.

Chapter 6 explores biodiversity and ecosystem services, assessing relationships between forest conditions and service provision using Bankura District as a case study. Chapter 7 examines livelihood dependencies and resource utilization patterns in Binpur-II block of Jhargram District, revealing spatial inequalities and the poverty-environment trap. Chapter 8 evaluates Joint Forest Management, analyzing the pioneering Arabari case to assess both ecological success and persistent social equity challenges. Chapter 9 addresses eco-development and sustainable forest management approaches, with emphasis on climate adaptation in the Sundarbans. The concluding chapter synthesizes findings and proposes pathways toward integrated social-ecological governance.

Throughout, the volume demonstrates how geographic information systems, remote sensing, and spatial analysis integrated with socio-economic investigation and participatory research enable evidence-based understanding of complex social-ecological dynamics. By combining scientific rigor with attention to social justice concerns, it contributes to both academic scholarship and practical efforts to achieve sustainable forest landscapes that balance ecological integrity with human dignity and welfare.

1.7 Significance and Contribution

This research makes several significant contributions to forest landscape science and sustainable development scholarship (Fig. 1.5). First, it provides comprehensive empirical documentation of forest change across multiple dimensions cover change, fragmentation, carbon dynamics, ecosystem services, and livelihood impacts in an understudied region. While West Bengal pioneered participatory forest governance through Joint Forest Management, systematic assessment of long-term outcomes remains limited. This volume addresses that gap through rigorous multi-method analysis.

Second, the research demonstrates how advanced geospatial technologies including high-resolution satellite imagery, machine learning classification algorithms, and spatial modeling tools enable precise monitoring and prediction of forest landscape dynamics. The validated models developed herein offer transferable methodological frameworks applicable to other regions facing similar pressures. Particularly valuable are the fragmentation assessment tools, carbon estimation models, and ecosystem service valuation approaches that can guide policy and planning.

Third, the volume's integrated social-ecological systems perspective moves beyond disciplinary silos to reveal crucial connections between biophysical processes and social dynamics. By examining not only ecological changes but also livelihood dependencies, power relations, gender dimensions, and institutional functioning, it illuminates how forest outcomes emerge from complex interactions between natural and social factors. This integration proves essential for designing interventions that are both ecologically effective and socially just.

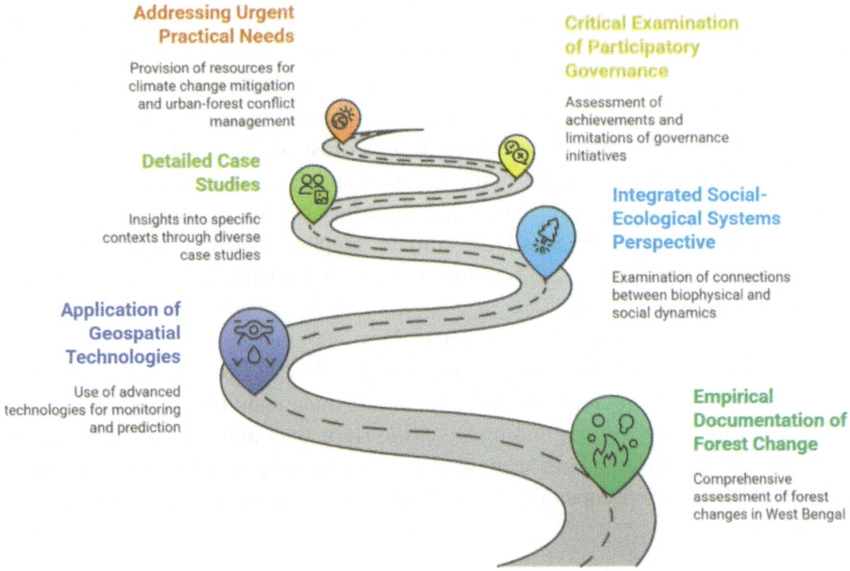

Fig. 1.5 Symmetrical diagram of research contributions to forest landscape science

Fourth, detailed case studies from diverse locations Jhargram, Bankura, Binpur-II, Siliguri, Arabari, Sundarbans provide granular insights into how general processes play out in specific contexts. These cases reveal heterogeneity often masked in aggregate analyses, showing how spatial location, community characteristics, historical trajectories, and institutional arrangements mediate forest outcomes. Such contextual understanding is crucial for policy design that avoids one-size-fits-all prescriptions.

Fifth, the critical examination of participatory governance initiatives including Joint Forest Management and the Forest Rights Act provides honest assessment of both achievements and persistent limitations. By documenting elite capture, gender exclusion, bureaucratic dominance, and incomplete devolution alongside genuine ecological recovery and livelihood improvements, the research offers nuanced understanding that can inform governance reforms. The identification of specific factors enabling or constraining participation provides actionable guidance for strengthening community-based conservation.

Finally, the research addresses urgent practical needs. As climate change intensifies, forests' roles in mitigation and adaptation become ever more critical. Understanding carbon storage patterns, identifying high-priority conservation areas, and designing climate-resilient management strategies require precisely the kind of evidence this volume provides. Similarly, as urban expansion continues, tools for assessing and managing urban-forest conflicts prove increasingly essential. The methodologies and findings presented herein offer concrete resources for policy-makers, forest managers, and civil society organizations working toward sustainable forest futures.

1.8 Conclusion

Forests represent far more than collections of trees they are complex social-ecological systems where natural processes and human societies continuously interact, shaping outcomes that profoundly affect both planetary ecological stability and human well-being. West Bengal's forest landscapes exemplify the challenges and opportunities inherent in reconciling conservation imperatives with development needs and social justice concerns. Through systematic analysis spanning historical transformations, contemporary ecological changes, livelihood dependencies, and governance innovations, this volume illuminates pathways toward sustainable forest futures.

The chapters that follow document both sobering realities and grounds for hope. Forest cover has declined dramatically, fragmentation threatens biodiversity and ecosystem functions, carbon stocks face ongoing pressures, and urban expansion generates intensifying conflicts. Simultaneously, participatory initiatives have achieved genuine ecological recovery in degraded areas, communities retain motivation and knowledge for sustainable stewardship, and technological innovations enable unprecedented monitoring and planning capacity.

Realizing sustainable forest landscapes requires fundamental shifts in governance from centralized control toward genuine power-sharing with local communities, from short-term exploitation toward long-term stewardship, from sectoral management toward integrated landscape approaches, and from top-down expert prescriptions toward knowledge co-production that respects both scientific and traditional wisdom. The evidence and analysis presented in subsequent chapters aims to contribute toward these essential transformations, offering insights relevant not only for West Bengal but for forest landscapes worldwide facing similar pressures and opportunities.

References

Agnoletti M (2007) Traditional forest knowledge: challenges and opportunities. For Ecol Manage 29:1–4

Agrawal A, Gibson CC (1999) Enchantment and disenchantment: the role of community in natural resource conservation. World Dev 27(4):629–649

Allen CD, Macalady AK, Chenchouni H, Bachelet D, McDowell N, Vennetier M et al (2010) A global overview of drought and heat-induced tree mortality reveals emerging climate change risks for forests. For Ecol Manage 259(4):660–684

Alongi DM (2008) Mangrove forests: resilience, protection from tsunamis, and responses to global climate change. Estuar Coast Shelf Sci 76(1):1–13

Andereg WR, Trugman AT, Badgley G, Anderson CM, Bartuska A, Ciais P et al (2020) Climate-driven risks to the climate mitigation potential of forests. Science 368(6497):eaaz7005

Armitage D, De Loë R, Plummer R (2012) Environmental governance and its implications for conservation practice. Conserv Lett 5(4):245–255

Arora NK, Mishra I (2019) United Nations Sustainable Development Goals 2030 and environmental sustainability: race against time. Environ Sustain 2(4):339–342

Assessment ME (2005) Ecosystems and human well-being: wetlands and water. World Resources Institute
Baccini A, Walker W, Carvalho L, Farina M, Sulla-Menashe D, Houghton RA (2017) Tropical forests are a net carbon source based on aboveground measurements of gain and loss. Science 358(6360):230–234
Bandyopadhyay S, Kar NS, Das S, Sen J (2014) River systems and water resources of West Bengal: a review. Geol Soc India Spec Publ 3(2014):63–84
Barlow J, Lennox GD, Ferreira J, Berenguer E, Lees AC, Nally RM et al (2016) Anthropogenic disturbance in tropical forests can double biodiversity loss from deforestation. Nature 535(7610):144–147
Barrington L, Ghosh S, Greene M, Har-Noy S, Berger J, Gill S et al (2011) Crowdsourcing earthquake damage assessment using remote sensing imagery. Ann Geophys 54(6)
Berkes F, Folke C (1998) Linking social and ecological systems for resilience and sustainability. Linking Soc Ecol Syst 1(4):4
Bisht S, Singh H, Kumar A (2024) Ecosystem services and maintaining. In: Urban forests, climate change and environmental pollution: physio-biochemical and molecular perspectives to enhance urban resilience, vol 447
Brown G, Kyttä M (2018) Key issues and priorities in participatory mapping: Toward integration or increased specialization? Appl Geogr 95:1–8
Costanza R, d'Arge R, De Groot R, Farber S, Grasso M, Hannon B et al (1997) The value of the world's ecosystem services and natural capital. Nature 387(6630):253–260
Curtis PG, Slay CM, Harris NL, Tyukavina A, Hansen MC (2018) Classifying drivers of global forest loss. Science 361(6407):1108–1111
Das GK (2021) Forest stands-case studies. Forests and forestry of West Bengal: survey and analysis. Springer International Publishing, Cham, pp 85–112
Dey D, Arunachalam A (2024) Sustainable and traditional environmental conservation in Asia
Díaz S, Pascual U, Stenseke M, Martín-López B, Watson RT, Molnár Z et al (2018) Assessing nature's contributions to people. Science 359(6373):270–272
Dietz T, Ostrom E, Stern PC (2003) The struggle to govern the commons. Science 302(5652):1907–1912
Domínguez L, Luoma C (2020) Decolonising conservation policy: how colonial land and conservation ideologies persist and perpetuate indigenous injustices at the expense of the environment. Land 9(3):65
Dutta S, Guchhait SK (2022) Assessment of land use land cover dynamics and urban growth of Kanksa Block in Paschim Barddhaman District, West Bengal. GeoJournal 87(2):971–990
Fahrig L (2017) Ecological responses to habitat fragmentation per se. Annu Rev Ecol Evol Syst 48:1–23
Fairhead J, Leach M (1996) Misreading the African landscape: society and ecology in a forest-savanna mosaic (no 90). Cambridge University Press
FAO (2020) Food and agriculture organization. Global forest resources assessment 2020—key findings. Rome. https://doi.org/10.4060/ca8753en. Retrieved 03 Nov 2022 from https://www.fao.org/forest-resources-assessment/2020/en/
Fischer J, Lindenmayer DB (2007) Landscape modification and habitat fragmentation: a synthesis. Glob Ecol Biogeogr 16(3):265–280
Folke C, Carpenter SR, Walker B, Scheffer M, Chapin T, Rockström J (2010) Resilience thinking: integrating resilience, adaptability and transformability. Ecol Soc 15(4)
Folke C, Colding J, Berkes F (2003) Building resilience and adaptive capacity in social-ecological systems. In: Navigating social-ecological systems. Cambridge University Press, Cambridge, UK, pp 352–387
Forest Survey of India (2021) India state of forest report 2021. Ministry of Environment, Forest and Climate Change, Government of India. https://fsi.nic.in/isfr2021/
Gautam A (2020) Forest management policies and tribals of Jharkhand: retrospect and prospect. Antrocom: Online J Anthropol 16(1)

Ghosh K, Ray M, Adak A, Halder SK, Das A, Jana A et al (2015) Role of probiotic Lactobacillus fermentum KKL1 in the preparation of a rice based fermented beverage. Biores Technol 188:161–168

Giri C, Ochieng E, Tieszen LL, Zhu Z, Singh A, Loveland T et al (2011) Status and distribution of mangrove forests of the world using earth observation satellite data. Glob Ecol Biogeogr 20(1):154–159

Gorelick N, Hancher M, Dixon M, Ilyushchenko S, Thau D, Moore R (2017) Google earth engine: planetary-scale geospatial analysis for everyone. Remote Sens Environ 202:18–27

Grimm NB, Faeth SH, Golubiewski NE, Redman CL, Wu J, Bai X, Briggs JM (2008) Global change and the ecology of cities. Science 319(5864):756–760

Griscom BW, Adams J, Ellis PW, Houghton RA, Lomax G, Miteva DA et al (2017) Natural climate solutions. Proc Natl Acad Sci 114(44):11645–11650

Gross E, Jayasinghe N, Brooks A, Polet G, Wadhwa R, Hilderink-Koopmans F (2021) A future for all: the need for human-wildlife coexistence. In: WWF, Gland, Switzerland). Design and infographics by Levent Köseoglu, WWF-Netherlands Text editing by ProofreadNOW. com Cover photograph: DNPWC-WWF Nepal, 3

Guha R (2000) The unquiet woods: ecological change and peasant resistance in the Himalaya. University of California Press

Haddad NM, Brudvig LA, Clobert J, Davies KF, Gonzalez A, Holt RD et al (2015) Habitat fragmentation and its lasting impact on Earth's ecosystems. Sci Adv 1(2):e1500052

Holling CS (2001) Understanding the complexity of economic, ecological, and social systems. Ecosystems 4(5):390–405

Huete A, Didan K, Miura T, Rodriguez EP, Gao X, Ferreira LG (2002) Overview of the radiometric and biophysical performance of the MODIS vegetation indices. Remote Sens Environ 83(1–2):195–213

Kamra L (2016) The politics of counterinsurgency and statemaking in modern India (Doctoral dissertation. University of Oxford

Karthigeyan K, Ranjan V, Gogoi R, Dash SS, Yadav VK (2022) Endemic and threatened vascular plants of West Bengal. Botanical Survey of India, Ministry of Environment, Forest & Climate Change, Government of India

Kashwan P (2017) Inequality, democracy, and the environment: a cross-national analysis. Ecol Econ 131:139–151

Keevers TL (2019) Cross-validation is insufficient for model validation. In: Joint and operations analysis division. Defence Science and Technology Group, Victoria, Australia

Khosla R, Bhardwaj A (2019) Urbanization in the time of climate change: examining the response of Indian cities. Wiley Interdiscip Rev: Climate Change 10(1):e560

Kimengsi JN, Owusu R, Charmakar S, Manu G, Giessen L (2023) A global systematic review of forest management institutions: towards a new research agenda. Landscape Ecol 38(2):307–326

Kumar K, Kerr JM (2012) Democratic assertions: the making of India's recognition of Forest Rights Act. Dev Chang 43(3):751–771

Kumar R, Saikia P (2020) Forest resources of Jharkhand, Eastern India: socio-economic and bio-ecological perspectives. Socio-economic and eco-biological dimensions in resource use and conservation: strategies for sustainability. Springer International Publishing, Cham, pp 61–101

Kushwaha SPS, Hazarika R (2004) Assessment of habitat loss in Kameng and Sonitpur Elephant Reserves. Curr Sci 1447–1453

Lambin EF, Meyfroidt P (2011) Global land use change, economic globalization, and the looming land scarcity. Proc Natl Acad Sci 108(9):3465–3472

Lax J, Köthke M (2017) Livelihood strategies and forest product utilisation of rural households in Nepal. Small-Scale for 16(4):505–520

Ma L, Liu Y, Zhang X, Ye Y, Yin G, Johnson BA (2019) Deep learning in remote sensing applications: a meta-analysis and review. ISPRS J Photogramm Remote Sens 152:166–177

Martin JF, Roy ED, Diemont SA, Ferguson BG (2010) Traditional ecological knowledge (TEK): ideas, inspiration, and designs for ecological engineering. Ecol Eng 36(7):839–849

Mather AS (2007) Recent Asian forest transitions in relation to foresttransition theory. Int for Rev 9(1):491–502

Meraj G, Singh SK, Kanga S, Islam MN (2022) Modeling on comparison of ecosystem services concepts, tools, methods and their ecological-economic implications: a review. Model Earth Syst Environ 8(1):15–34

Mitra K (1995) Dynamics of forest product consumption: the role of economic development and forest policies in West Bengal, India. University of Florida

MoEFCC (2019) Proposed Indian Forest (Amendment) Act, 2019. Forest Policy Division, Ministry of Environment, Forests & Climate Change, Government of India, New Delhi

Mondal P, Southworth J (2010) Protection versus commercial management: spatial and temporal analysis of land cover changes in the tropical forests of Central India. For Ecol Manage 259(5):1009–1017

Mukhopadhyay B, Khan A (2015) A reevaluation of the snowmelt and glacial melt in river flows within Upper Indus Basin and its significance in a changing climate. J Hydrol 527:119–132

Nelson MK (2014) Indigenous science and traditional ecological knowledge: persistence in place. In: The world of indigenous North America. Routledge, pp 188–214

Nizamani MM, Zhang Q, Muhae-Ud-Din G, Awais M, Qayyum M, Farhan M et al (2024) Application of GIS and remote-sensing technology in ecosystem services and biodiversity conservation. In: Deep learning for multimedia processing applications. CRC Press, pp 284–321

Oldekop JA, Rasmussen LV, Agrawal A, Bebbington AJ, Meyfroidt P, Bengston DN et al (2020) Forest-linked livelihoods in a globalized world. Nat Plants 6(12):1400–1407

Ostrom E (2009) A general framework for analyzing sustainability of social-ecological systems. Science 325(5939):419–422

Ostrom E (2010) Beyond markets and states: polycentric governance of complex economic systems. Am Econ Rev 100(3):641–672

Ostrom E, Cox M (2010) Moving beyond panaceas: a multi-tiered diagnostic approach for social-ecological analysis. Environ Conserv 37(4):451–463

Pal S, Das S, Samanta S, Chattopadhyay S, Hazra S (2024) A preliminary appraisal of the mobility of tribal communities in the Sundarban Biosphere Reserve, India. SN Soc Sci 4(3):58

Pan Y, Birdsey RA, Fang J, Houghton R, Kauppi PE, Kurz WA et al (2011) A large and persistent carbon sink in the world's forests. Science 333(6045):988–993

Pandey S, Kumari N, Dash SK, Al Nawajish S (2022) Challenges and monitoring methods of forest management through geospatial application: a review. Adv Remote Sens For Monit 289–328

Pelletier C, Valero S, Inglada J, Champion N, Dedieu G (2016) Assessing the robustness of Random Forests to map land cover with high resolution satellite image time series over large areas. Remote Sens Environ 187:156–168

Peluso NL (1992) The political ecology of extraction and extractive reserves in East Kalimantan, Indonesia. Dev Change 23(4):49–74

Poffenberger M, McGean B (1996) Village voices, forest choices. In: Joint forest management in India Oxford University Press Delhi

Ram AK, Yadav NK, Kandel PN, Mondol S, Pandav B, Natarajan L et al (2021) Tracking forest loss and fragmentation between 1930 and 2020 in Asian elephant (Elephas maximus) range in Nepal. Sci Rep 11(1):19514

Ramachandra TV, Bharath S, Bharath AH (2020) Insights of forest dynamics for the regional ecological fragility assessment. J Indian Soc Remote Sens 48(8):1169–1189

Rawal RS, Bhatt ID, Sekar KC, Nandi SK (2013) The himalayan biodiversity: richness, representativeness, uniqueness and life-support values. GB Pant Institute of Himalayan Environment and Development (GBPIHED), Kosi-Katarmal, Almora, Uttarakhand, India, 84pp

Rejou-Mechain M, Muller-Landau HC, Detto M, Thomas SC, Le Toan T, Saatchi SS et al (2014) Local spatial structure of forest biomass and its consequences for remote sensing of carbon stocks. Biogeosciences 11(23):6827–6840

Richards JF, Flint EP (1994) A century of land-use change in South and Southeast Asia. In: Effects of land-use change on atmospheric CO_2 concentrations: South and Southeast Asia as a case study. Springer New York, New York, NY, pp 15–66
Ritter E, Dauksta D (2013) Human–forest relationships: ancient values in modern perspectives. Environ Dev Sustain 15(3):645–662
Roy A, Fleischman F (2022) The evolution of forest restoration in India: the journey from precolonial to India's 75th year of independence. Land Degrad Dev 33(10):1527–1540
Roy PS, Tomar S (2000) Biodiversity characterization at landscape level using geospatial modelling technique. Biol Cons 95(1):95–109
Sarkar S, Manna H, Roy SK, Dolui M, Hossain M (2024) Synergizing remote sensing and ecological indicators (RSEIs) for evaluating ecological environmental quality (EEQ) in Asansol Municipal Corporation: an integrated approach. Environ Monit Assess 196(7):631
Sen A (2021) A political ecology of forest conservation in India: communities, wildlife and the state. Routledge
Seto KC, Güneralp B, Hutyra LR (2012) Global forecasts of urban expansion to 2030 and direct impacts on biodiversity and carbon pools. Proc Natl Acad Sci 109(40):16083–16088
Shoko J, Muboko N, Gandiwa E (2024) Focus group discussion method as a tool for eliciting local community views for decision making; a case of the save valley conservancy in the Southeast Lowveld of Zimbabwe
Sivaramakrishnan K (1999) Modern forests: statemaking and environmental change in colonial Eastern India. Stanford University Press
Springate-Baginski O, Blaikie PIERS (2007) Forests people and power: the political ecology of reform in South Asia. Earthscan, London
Turner DP, Ritts WD, Cohen WB, Gower ST, Zhao M, Running SW et al (2003) Scaling gross primary production (GPP) over boreal and deciduous forest landscapes in support of MODIS GPP product validation. Remote Sens Environ 88(3):256–270
Walker B, Holling CS, Carpenter SR, Kinzig A (2004) Resilience, adaptability and transformability in social–ecological systems. Ecol Soc 9(2)
Wang X, Blanchet FG, Koper N (2014) Measuring habitat fragmentation: an evaluation of landscape pattern metrics. Methods Ecol Evol 5(7):634–646
Woodcock CE, Allen R, Anderson M, Belward A, Bindschadler R, Cohen W et al (2008) Free access to Landsat imagery. Science 320:1011
Wulder MA, Coops NC, Roy DP, White JC, Hermosilla T (2018) Land cover 2.0. Int J Remote Sens 39(12):4254–4284
Xu D, Wang H, Xu W, Luan Z, Xu X (2021) LiDAR applications to estimate forest biomass at individual tree scale: opportunities, challenges and future perspectives. Forests 12(5):550
Young OR (2002) The institutional dimensions of environmental change: fit, interplay, and scale. MIT Press

Chapter 2
Historical Transformation of Forest Cover: Colonial Era to Present Day

Abstract This chapter examines the historical transformation of forest cover in West Bengal from the pre-colonial era to the present day, analyzing how changing governance paradigms have reshaped both ecological systems and human-forest relationships. Prior to British colonization, sophisticated indigenous management systems maintained forest ecosystems through community-based stewardship, incorporating sacred groves, rotational harvesting, and traditional ecological knowledge that ensured sustainable resource utilization. The colonial period marked a fundamental disruption with the establishment of the Imperial Forest Department in 1864 and the implementation of scientific forestry regimes that reconceptualized forests primarily as economic resources. The Forest Acts of 1865 and 1878 introduced reserved forests and criminalized traditional practices, fundamentally altering social relationships with forest landscapes. This extractive approach reduced forest cover from approximately 30% in the early twentieth century to 13.4% by 1988. Post-independence forest governance initially continued colonial approaches before evolving toward participatory frameworks. West Bengal pioneered Joint Forest Management in the late 1980s, creating Forest Protection Committees and offering communities revenue-sharing arrangements. While this approach demonstrated potential for forest regeneration, implementation challenges and power imbalances have limited effectiveness. Contemporary deforestation patterns reveal complex spatial and temporal dynamics driven by agricultural conversion (67% of forest loss), settlement expansion, infrastructure development, and mining operations. The struggle for community rights remains central, with legislation like the Forest Rights Act of 2006 providing theoretical recognition but experiencing inconsistent implementation. This historical analysis reveals both persistent barriers to sustainable forest governance and potential pathways toward more equitable human-forest relationships, emphasizing the need for adaptive management approaches that integrate traditional knowledge with contemporary conservation challenges.

Keywords Historical transformation · Pre-colonial era · British colonization · Indigenous management · West Bengal

P. K. Shit and M. Hasanuzzaman, *Forest Landscape Dynamics in West Bengal*, Advances in Asian Human-Environmental Research, https://doi.org/10.1007/978-3-032-23700-2_2

2.1 Introduction

The forests of West Bengal represent a living chronicle of changing human–environment relationships, bearing witness to profound transformations across centuries of evolving governance regimes (Sen 2021). From indigenous stewardship to colonial exploitation and contemporary efforts at participatory management, these forest landscapes embody the complex interplay between ecology, politics, and cultural systems (Clay 2016). This chapter examines the historical trajectory of forest cover transformation in West Bengal from the pre-colonial era to the present day, highlighting the shifting paradigms of forest management and their ecological consequences.

Prior to British colonization, forests in West Bengal were managed through sophisticated indigenous systems that integrated ecological knowledge with cultural practices (Das Gupta 2014). Local communities maintained sacred groves, implemented rotational harvesting, and governed forest access through traditional institutions that ensured sustainable resource utilization (Kandari et al. 2014). This equilibrium was dramatically disrupted during the colonial period, when British authorities established the Imperial Forest Department in 1864 and implemented scientific forestry regimes that reconceptualized forests primarily as economic resources (Guha 1983; Sivaramakrishnan 1999). The introduction of reserved forests and criminalizing of traditional practices fundamentally altered both ecological systems and social relationships with forest landscapes.

The post-independence era has witnessed multiple policy shifts, from continued state control to innovative participatory approaches like Joint Forest Management, which emerged in West Bengal during the late 1980s as a response to widespread deforestation and community resistance (Poffenberger 1990; Bhattacharya et al. 2010). These evolving governance frameworks have occurred against a backdrop of significant land-use changes, with forest cover declining from approximately 30% in the early twentieth century to 13.4% by 1988, shaped by factors including agricultural expansion, infrastructure development, and settlement growth (Roy and Tomar 2001; Mondal et al. 2016).

Throughout these transitions, the question of community rights and access has remained central to forest politics in West Bengal (Banerjee et al. 2010). Indigenous and forest-dependent communities have experienced cycles of dispossession, resistance, and partial recognition, with recent legislation like the Forest Rights Act of 2006 representing important but incompletely realized opportunities for rights restoration (Sarin and Springate-Baginski 2010). As West Bengal now confronts new challenges including climate change, urbanization pressures, and evolving market demands for forest products, understanding this historical context becomes essential for crafting sustainable and equitable forest governance for the future.

This chapter traces these interconnected historical processes, examining how changing management paradigms have transformed both the physical landscape and social relationships with forests in West Bengal. By analyzing the legacies of colonial forestry, post-independence policy evolution, patterns of deforestation, and the struggle for community rights, we gain critical insights into the complex forces that continue to shape forest governance in the region today.

2.2 Pre-colonial Forest Management Systems

Prior to British colonization, the forests of West Bengal were governed through sophisticated indigenous management systems that reflected deep ecological knowledge and cultural integration (Fig. 2.1). Local communities, particularly tribal groups such as the Santhal, Munda, and Oraon, maintained sacred groves known as "sarnas" where ancient practices prohibited unauthorized felling of trees and hunting, effectively creating protected biodiversity zones (Gadgil and Vartak 1976; Malhotra and Bhattacharya 2010). These communities implemented rotational harvesting practices that allowed for forest regeneration, with collective decision-making structures determining when and which forest areas could be utilized for resources (Guha 1983).

The forest management was inextricably linked to religious and cultural practices, with village elders and spiritual leaders enforcing customary laws that dictated sustainable extraction methods (Sivaramakrishnan 1999). These systems were characterized by community-owned forest tracts where usufruct rights were carefully allocated based on need rather than commercial value, creating a balanced ecosystem approach (Rangarajan 1999). Notably, these management regimes incorporated traditional ecological knowledge regarding seasonal variations, medicinal properties of plants, and habitat preservation for wildlife (Berkes et al. 1994). The pre-colonial period also featured "gram sabhas" (village councils) that mediated conflicts regarding forest access and resource allocation, ensuring equitable distribution across different social groups (Sundar 2000). Unlike the subsequent colonial forestry paradigm that commodified timber, these indigenous systems viewed forests holistically as sources of food security, medicinal resources, spiritual significance, and ecological resilience (Guha and Gadgil 1989). Even feudal rulers recognized the

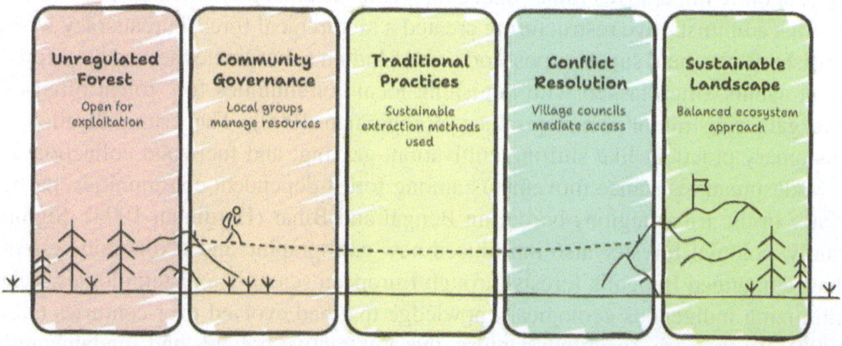

Fig. 2.1 Symmetrical diagram illustrating the transition of indigenous forest management systems

importance of these community-managed systems, often formalizing local control through royal decrees that acknowledged traditional boundaries and management practices (Bandopadhyay 2010). This nuanced understanding of forest ecosystems developed over centuries ultimately created sustainable landscapes where human settlements and natural forests coexisted in relative equilibrium (Schlich 1907; Rammer and Seidl 2015).

2.3 British Colonial Forestry Practices

The British colonial intervention in West Bengal's forests marked a fundamental shift from community-based management to state-controlled exploitation, beginning with the establishment of the Imperial Forest Department in 1864 (Fig. 2.2) (Guha 1983). This transformation was catalyzed by the growing timber demands for shipbuilding, railway expansion, and military infrastructure, which positioned forests primarily as revenue-generating resources rather than ecological or cultural entities (Barton 2000; Stebbing 1925). The Forest Act of 1865, subsequently revised in 1878, institutionalized this paradigm shift by introducing the novel concept of "reserved forests," effectively criminalizing traditional access rights of indigenous communities while centralizing control under colonial authorities (Gadgil and Guha 1993). In West Bengal specifically, the commercially valuable sal (*Shorea robusta*) forests of the Midnapore, Bankura, and Purulia Districts were systematically surveyed, demarcated, and placed under scientific forestry regimes that prioritized monoculture plantations over mixed native vegetation (Sivaramakrishnan 1999). Colonial foresters like Dietrich Brandis introduced European silvicultural techniques ill-suited to Bengal's subtropical ecosystems, implementing clear-felling methods and emphasizing fast-growing exotic species like teak (*Tectona grandis*) and eucalyptus that dramatically altered forest composition (Grove 1996; Rajan 2006). The Working Plan system, introduced in the late nineteenth century, compartmentalized forests into management units where harvesting cycles and yield calculations were determined by economic imperatives rather than ecological sustainability (Schlich 1907).

This administrative restructuring created a hierarchical forest bureaucracy where British officers held superior positions while Indian subordinates served as enforcement agents, often tasked with punishing local communities for "forest offenses" like gathering minor forest produce (Rangarajan 1999). The criminalization of customary practices like shifting cultivation, grazing, and fuelwood collection led to widespread resistance movements among forest-dependent communities, particularly in the tribal regions bordering Bengal and Bihar (Hardiman 1994). Significantly, colonial forestry also introduced new cartographic and taxonomic systems that reimagined Bengal's forests through European scientific epistemologies, delegitimizing indigenous ecological knowledge that had evolved over centuries (Kerr 1999). By the early twentieth century, these extractive policies had fundamentally transformed West Bengal's Forest landscapes, reducing dense, biodiverse forests to

Fig. 2.2 Symmetrical diagram presenting colonial impact on West Bengal Forest

simplified timber production systems while simultaneously creating new categories of environmental crime and marginalized forest dwellers (Sen 2021).

2.4 Post-independence Forest Policy Changes

The post-independence era witnessed significant transformations in West Bengal's Forest governance, characterized by an initial continuation of colonial exploitation followed by gradual shifts toward community participation and ecological sensitivity (Banerjee et al. 2010) (Fig. 2.3). The immediate post-independence period (1947–1970) largely maintained colonial forestry structures with the Indian Forest Act of 1927 remaining operational, though the state's forests were increasingly exploited to meet reconstruction and industrialization demands (Guha 1983). The nationalization of forests under the West Bengal Estate Acquisition Act of 1953 transferred ownership of private forests to the state government, centralizing control while further marginalizing forest-dependent communities that had already been dispossessed during colonial rule (Banerjee 1997; Sivaramakrishnan 1999). A significant policy shift emerged in the 1970s with growing recognition of deforestation's consequences, leading to the Wildlife Protection Act (1972) and Forest Conservation Act (1980), which prioritized conservation but often through exclusionary approaches that restricted community access (Gadgil and Guha 1993). The most revolutionary transformation came in the late 1980s when West Bengal pioneered the Joint Forest Management (JFM) program, beginning experimentally in Arabari forest range of Midnapore District in 1972 but formalized through a government resolution in 1989, which reconceptualized forest-community relationships by establishing Forest Protection Committees (FPCs) and offering participating communities a 25% share of timber revenues and access to non-timber forest products (Poffenberger 1990; Chopra and Chatterjee 1990).

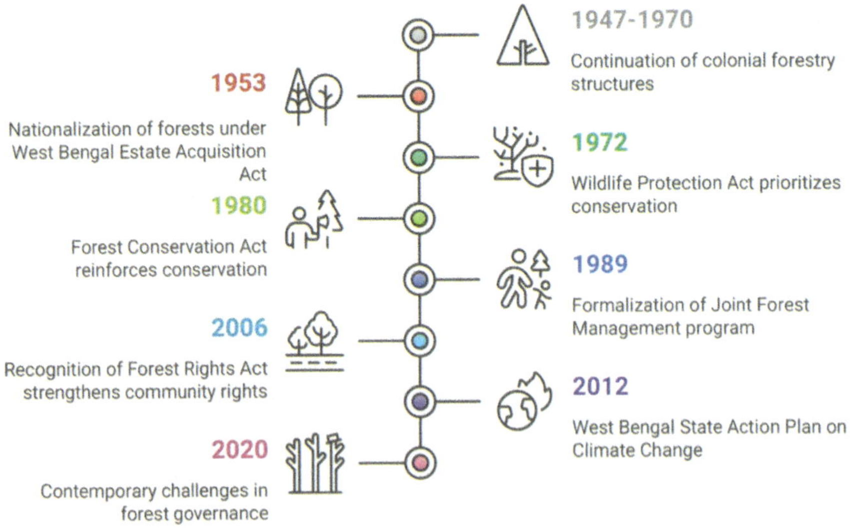

Fig. 2.3 Schematic representation of the evolution of forest policy in West Bengal

This participatory approach emerged in response to extensive forest degradation and local resistance movements, with subsequent research documenting remarkable regeneration of previously degraded sal forests and improved rural livelihoods across southwestern districts like Bankura, Purulia, and Midnapore (Saxena 1997; Bhattacharya et al. 2010). The 1988 National Forest Policy further reinforced this paradigm shift by acknowledging forest dwellers' needs and traditional knowledge, though implementation often remained constrained by bureaucratic structures and power asymmetries (Sundar 2001). Recent decades have seen additional policy layers emerge, including the Scheduled Tribes and Other Traditional Forest Dwellers (Recognition of Forest Rights) Act of 2006, which theoretically strengthened community rights but has experienced implementation challenges in West Bengal due to competing conservation agendas and political complexities (Das 2022). Additionally, climate change concerns have introduced new dimensions to forest policy, with West Bengal's State Action Plan on Climate Change (2012) and Green Bengal Mission emphasizing carbon sequestration alongside biodiversity conservation and livelihood security (Khan 1997; Choudhury and Roy 2021). Contemporary challenges include reconciling conservation objectives with development pressures, addressing continued forest fragmentation, and navigating tensions between centralized and decentralized governance structures that have characterized West Bengal's Forest policy evolution since independence (Bhattacharya 2014).

2.5 Land-Use Changes and Deforestation Patterns

West Bengal has experienced profound forest landscape transformations over the past century, with distinct spatial and temporal patterns of deforestation driven by multiple socio-economic and political factors (Chakraborty et al. 2024). Historically, forest cover in the state declined from approximately 30% in the early twentieth century to merely 13.4% by 1988, with the steepest rates of deforestation occurring during the refugee resettlement period following the 1947 partition and the Green Revolution intensification of the 1960s–70s (Roy and Tomar 2001; Mondal et al. 2016). Remote sensing analyses reveal heterogeneous deforestation patterns across the state's diverse physiographic zones: the southwestern lateritic region (comprising Purulia, Bankura, and West Midnapore Districts) experienced fragmentation primarily through fuelwood extraction and shifting cultivation until the 1980s, while the northern Himalayan Districts witnessed deforestation along steep slopes due to road construction, hydroelectric projects, and expansion of tea plantations (Sudhakar Reddy et al. 2016; Baral et al. 2012). The Sundarbans mangrove forests in the south have faced a different trajectory, with approximately 1.2% of mangrove area lost annually between 1975 and 2005 due to aquaculture expansion, agricultural encroachment, and rising sea levels a pattern exacerbated by climate change vulnerabilities (Giri et al. 2007; Ghosh et al. 2015). Land-use change analysis utilizing multi-temporal satellite imagery demonstrates that agricultural conversion has been the predominant driver of deforestation, accounting for 67% of forest loss between

1990 and 2010, followed by settlement expansion (18%), infrastructure development (8%), and commercial plantations (7%) (Akubia and Bruns 2019; Chakraborty et al. 2018; Kumar et al. 2014) (Fig. 2.4). Notably, districts bordering Jharkhand and Odisha have exhibited distinctive conversion patterns where mining operations and industrial development have created deforestation hotspots with annual forest loss rates exceeding 2.5% in certain micro-regions (Chakraborty et al. 2018). The spatial configuration of deforestation has also evolved over time, while early deforestation occurred primarily along forest peripheries, contemporary forest loss (2000–2020) reveals more complex patterns of perforation and attrition within forest interiors, suggesting changing accessibility dynamics and enforcement challenges (Nandy et al. 2011). Crucially, temporal analyses reveal that deforestation rates declined significantly after 1990 in districts with successful Joint Forest Management implementation, though these gains have been partially offset by qualitative degradation where native species diversity diminished even as canopy cover increased through monoculture plantations (Bhattacharya et al. 2010; Gorain et al. 2025).

Recent studies employing machine learning algorithms to integrate socio-economic and biophysical variables have identified proximity to roads, population density, and poverty rates as the strongest predictors of continued deforestation risk, with these factors explaining 73% of observed forest conversion patterns (Meiyappan

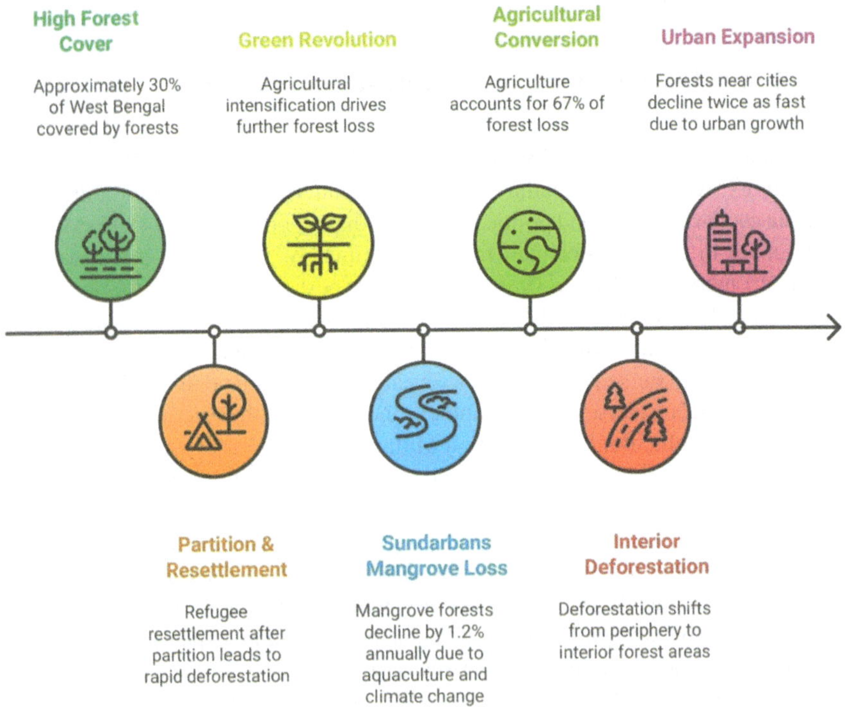

Fig. 2.4 Diagram illustrating the major deforestation events in West Bengal

et al. 2017; Thakur et al. 2021). Furthermore, land-use transition matrices indicate complex forest-agriculture-settlement conversion trajectories where approximately 32% of deforested areas undergo multiple land-use transitions within a decade, reflecting dynamic land markets and shifting economic opportunities (Kayal et al., 2025). Contemporary trends reveal emerging pressures from urban expansion and infrastructure corridors, with forest areas within 30 km of major cities declining at twice the rate of more remote forests, signaling new challenges for forest conservation in an urbanizing landscape (Mukhopadhyay et al. 2021; Sharma et al. 2024) (Fig. 2.5).

2.6 Evolution of Community Rights and Access of West Bengal

The evolution of community rights and access to forests in West Bengal represents a complex narrative of displacement, resistance, and gradual recognition spanning pre-colonial, colonial, and post-colonial periods (Fig. 2.6). Indigenous communities like the Santhal, Munda, Oraon, and Kheria traditionally exercised customary rights over forests through systems of common property management, with clearly defined access rules, seasonal extraction protocols, and cultural practices that regulated resource use prior to colonial intervention (Debnath 2011; Sen and Bhakat 2021) (Fig. 2.7). The British colonial regime fundamentally disrupted these systems through the doctrine of eminent domain, with legislation such as the Indian Forest Act of 1865 and its revised version of 1878 recategorizing forest lands as "state property" and redefining traditional practices as "concessions" rather than rights, resulting in widespread criminalization of customary forest use (Guha 1983; Sivaramakrishnan 1999). This dispossession triggered numerous resistance movements, including the Santhal Rebellion (1855–56) and the Chuar Uprising (1799), which explicitly contested forest access restrictions and represented early assertions of community rights (Dasgupta 1999; Rangarajan 1999).

Post-independence, the initial forest policy frameworks maintained colonial restrictions, with the West Bengal Private Forests Act of 1948 and the estate acquisition processes of the 1950s further marginalizing community access by prioritizing commercial timber extraction and industrial demand (Banerjee et al. 2010; Sen and Lalhrietpui 2006). A pivotal transition began in the 1970s with the experimental Joint Forest Management (JFM) initiative in Arabari Forest Range of Midnapore District, which recognized the failure of exclusionary conservation and created institutional mechanisms for community involvement through Forest Protection Committees (FPCs), representing the first formal recognition of community stake in forest management since colonial intervention (Joshi 1983; Roy 1992).

The subsequent formalization of JFM through government orders in 1989 and the West Bengal Forest Rules expanded this approach, granting participating communities 25% of timber revenues and regulated access to non-timber forest products

Fig. 2.5 Representative field photographs from the study area showing different aspects of forest landscape dynamics: **a** active forest fire affecting the understory vegetation; **b** post-fire burnt forest floor and degraded tree stand; **c** forest road passing through a sal-dominated woodland; **d** gully erosion and associated land degradation near the forest fringe; **e** patchy scrub and grassland indicating vegetation degradation; and **f** plantation/managed forest plot showing human intervention in forest restoration

(NTFPs), though decision-making authority remained asymmetrically concentrated with the Forest Department (Poffenberger and McGean 1996; Bhattacharya et al. 2010). The effectiveness of these arrangements varied significantly, while some regions like Bankura and Midnapore documented improved forest conditions and livelihood outcomes, persistent challenges included elite capture of benefits, gender disparities in committee representation, and conflicts over share distribution (Debnath 2023; Ghosal 2011). The passage of the Scheduled Tribes and Other Traditional

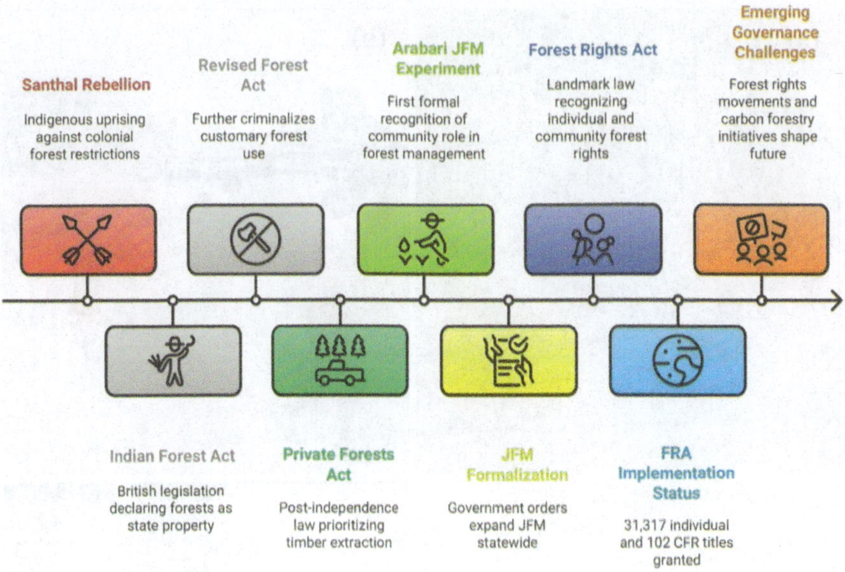

Fig. 2.6 Symmetrical diagram depicting evolution of community rights and access to forests in West Bengal

Forest Dwellers (Recognition of Forest Rights) Act in 2006 (FRA) potentially represented a paradigm shift by legally recognizing individual and community forest rights, including the crucial provision for Community Forest Resource (CFR) rights that empower gram sabhas with forest governance authority (Sarin and Springate-Baginski 2010). However, implementation in West Bengal has been notably inconsistent, while 31,317 individual forest rights titles were distributed by 2020, community rights recognition has lagged significantly with only 102 CFR titles granted statewide, concentrated primarily in Jalpaiguri and Alipurduar Districts (Dash and Behera 2018; Ghosh et al. 2019). Analysis of this implementation gap reveals institutional resistance from the Forest Department, procedural barriers in claim processing, and competing conservation priorities as limiting factors (Kumar et al. 2015; Lee and Wolf 2018). Recent developments include the emergence of forest rights movements like the Jungle Mahal Adivasi Adhikar Manch and increased activism demanding implementation of community forest rights, alongside new governance challenges arising from market linkages for forest products and emerging carbon forestry initiatives like REDD + that introduce complex questions about community benefit-sharing and access rights in an era of climate change response (McDermott et al. 2024; Rana 2016).

Fig. 2.7 **a** Indigenous women and children walking through a sal-dominated forest while collecting dry fuelwood, illustrating everyday dependence on forest resources. **b** Local tribal women transporting heavy bundles of collected firewood on their heads across an open dry landscape, reflecting subsistence fuelwood extraction practices. **c** A bicycle loaded with a large stack of fuelwood bundles, demonstrating small-scale forest biomass transport for household use or local sale. **d** A rural resident carrying a sack of freshly cut grass on a bicycle within a plantation landscape, indicating the collection of fodder and non-timber forest resources for livelihood support

2.7 Conclusion

The forests of West Bengal have undergone profound transformations shaped by shifting governance paradigms across pre-colonial, colonial, and contemporary periods. Indigenous stewardship practices, which sustained biodiverse ecosystems through community-based management systems, were systematically undermined during the colonial era when British authorities imposed scientific forestry regimes that prioritized economic exploitation over ecological sustainability. This colonial legacy established patterns of centralized control that continued to influence forest governance after independence. Post-independence forest policy initially maintained extractive approaches before gradually evolving toward more participatory frameworks. The pioneering Joint Forest Management program demonstrated the potential for forest regeneration through community collaboration, though implementation challenges and power imbalances have limited its effectiveness. Throughout this period, deforestation patterns have been shaped by various socio-economic forces from refugee resettlement after Partition to agricultural intensification during the Green Revolution and recent urbanization pressures. A central theme in this history is the ongoing struggle for community rights and access. Despite legislation like the Forest Rights Act of 2006, recognition of community forest resource rights remains inconsistent, reflecting tensions between conservation goals, development priorities, and social justice claims. Looking ahead, sustainable forest management in West Bengal requires integrating traditional ecological knowledge, addressing power asymmetries in participatory structures, and developing adaptive responses to challenges like climate change. Success will depend on reconciling multiple objectives through inclusive institutional arrangements that learn from historical experiences. This historical analysis reveals both persistent barriers to sustainable forest governance and potential pathways toward more equitable human-forest relationships. The story of West Bengal's forests ultimately represents a continuous process of adaptation and resilience in the face of profound ecological and social change.

References

Akubia JE, Bruns A (2019) Unravelling the frontiers of urban growth: spatio-temporal dynamics of land-use change and urban expansion in greater Accra metropolitan area, Ghana. Land 8(9):131

Bandopadhyay A (2010) The colonial legacy of forest policies in India. Soc Sci 38(1/2):53–76

Banerjee AK (1997) Decentralization and devolution of forest management in Asia and the Pacific

Banerjee A, Ghosh S, Springate-Baginski O (2010) The creation of West Bengal's forest underclass: an historical institutional analysis of forest rights deprivations. IPPG Discuss Papers 51:1–26

Baral NR, Acharya DP, Rana CJ (2012) Study on drivers of deforestation and degradation of forests in high mountain regions of Nepal. In: REDD cell. Ministry of Forest and Soil Conservation, Nepal

Barton G (2000) Keepers of the Jungle: environmental management in British India, 1855–1900. Historian 62(3):557–574

Berkes F, Folke C, Gadgil M (1994) Traditional ecological knowledge, biodiversity, resilience and sustainability. In: Biodiversity conservation: problems and policies. Papers from the biodiversity programme Beijer international Institute of ecological economics royal Swedish academy of sciences. Springer Netherlands, Dordrecht, pp 269–287

Bhattacharya S (2014) Forest and biodiversity conservation in ancient Indian culture: a review based on old texts and archaeological evidences. Int Lett Soc Human Sci 19:35–46

Bhattacharya P, Pradhan L, Yadav G (2010) Joint forest management in India: experiences of two decades. Resour Conserv Recycl 54(8):469–480

Chakraborty A, Seshasai MVR, Reddy CS, Dadhwal VK (2018) Persistent negative changes in seasonal greenness over different forest types of India using MODIS time series NDVI data (2001–2014). Ecol Ind 85:887–903

Chakraborty A, Datta D, Roy AK (2024) Spatial patterns of forest fragmentation and human–elephant conflicts in south-western West Bengal, India: a multitemporal assessment. In: Vegetation dynamics and crop stress. Academic Press, pp 63–91

Chopra S, Chatterjee D (1990) Integrating conservation and development: a case study of the socio-economic forestry complex at Arabari, West Bengal. Int Tree Crops J 6(2–3):193–204

Choudhury M, Roy P (2021) Climate change and lagging conviction. Voice Environ Newsl 4(01)

Clay N (2016) Producing hybrid forests in the Congo Basin: a political ecology of the landscape approach to conservation. Geoforum 76:130–141

Das BK (2022) How to make forest governance reform real? Formal and informal institutions during implementation of the forest rights act 2006 in India. Environ Dev 43:100729

Das Gupta S (2014) Colonial rule versus indigenous knowledge in Bengal's western frontier. RCC Perspect 3:45–52

Dasgupta A (1999) Ethnic problems and movements for autonomy in Darjeeling. Soc Sci 47–68

Debnath M (2023) Reclaiming indigenous voice and knowledge in the era of (re) colonization, insights from a rural indigenous Santal community. Fourth World J 23(1):84–108

Debnath MK (2011) Living on the edge: the predicament of a rural indigenous Santal community in Bangladesh. Doctoral dissertation

Gadgil M, Guha R (1993) This fissured land: an ecological history of India. University of California Press

Gadgil M, Vartak VD (1976) The sacred groves of Western Ghats in India. Econ Bot 30(2):152–160

Ghosal S (2011) Pre-colonial and colonial forest culture in the Presidency of Bengal. Hum Geogr: J Stud Res Hum Geogr 5(1)

Ghosh S, Das N, Das I, Maulik U (2019) Understanding deep learning techniques for image segmentation. ACM Comput Surv (CSUR) 52(4):1–35

Ghosh S, Bakshi M, Bhattacharyya S, Nath B, Chaudhuri P (2015) A review of threats and vulnerabilities to mangrove habitats: with special emphasis on east coast of India

Giri C, Pengra B, Zhu Z, Singh A, Tieszen LL (2007) Monitoring mangrove forest dynamics of the Sundarbans in Bangladesh and India using multi-temporal satellite data from 1973 to 2000. Estuar Coast Shelf Sci 73(1–2):91–100

Gorain S, Malakar A, Dutta S, Das S (2025) Assessing forest cover, carbon stocks and fire dynamics in India and economic valuation of forest carbon in Asia: a meta-analysis. Discov for 1(1):16

Grove RH (1996) Green imperialism: colonial expansion, tropical island Edens and the origins of environmentalism, 1600–1860. Cambridge University Press

Guha R, Gadgil M (1989) State forestry and social conflict in British India. Past Present 123:141–177

Guha R (1983) Forestry in British and post-British India: a historical analysis. Econ Political Wkly 1882–1896

Hardiman D (1994) Power in the forest: the Dangs 1820–1920

Joshi G (1983) Forests and forest policy in India. Soc Sci 43–52

Kandari LS, Bisht VK, Bhardwaj M, Thakur AK (2014) Conservation and management of sacred groves, myths and beliefs of tribal communities: a case study from north-India. Environ Syst Res 3(1):16

Kayal P, Das S, Chowdhury IR (2025) Modeling agricultural land suitability using MCDM-AHP techniques in semi-arid region of West Bengal, India. In: Surface, sub-surface hydrology and management: application of geospatial and geostatistical techniques. Springer Nature Switzerland, Cham, pp 657–694

Kerr IJ (1999) Hybrid histories: forests, frontiers and wildness in Western India. In: Skaria A. Oxford University Press, Delhi, 324 pp

Khan MR (1997) Social dimension of sustainable development: an inquiry into the forest sectors of Bangladesh and West Bengal, India. University of Maryland, College Park

Kumar R, Nandy S, Agarwal R, Kushwaha SPS (2014) Forest cover dynamics analysis and prediction modeling using logistic regression model. Ecol Ind 45:444–455
Kumar K, Singh NM, Kerr JM (2015) Decentralisation and democratic forest reforms in India: moving to a rights-based approach. Forest Policy Econ 51:1–8
Lee JI, Wolf SA (2018) Critical assessment of implementation of the Forest Rights Act of India. Land Use Policy 79:834–844
Malhotra KC, Bhattacharya P (2010) Forest and livelihood. CESS, Hyderabad, p 246
McDermott C, Burns SL, Brockhaus M, Bong IW, Hafferty C, Hirons M et al (2024) A political ecology and economy of key trends in international forest governance. In: International forest governance: a critical review of trends, drawbacks, and new approaches. A global assessment report. IUFRO, Vienna, pp 19–56
Meiyappan P, Roy PS, Sharma Y, Ramachandran RM, Joshi PK, DeFries RS, Jain AK (2017) Dynamics and determinants of land change in India: integrating satellite data with village socioeconomics. Reg Environ Change 17(3):753–766
Mondal MS, Sharma N, Garg PK, Kappas M (2016) Statistical independence test and validation of CA Markov land use land cover (LULC) prediction results. Egypt J Remote Sens Space Sci 19(2):259–272
Mukhopadhyay A, Mitra D, Hazra S (eds) (2021) Sundarbans Mangrove systems: a geo-informatics approach. CRC Press
Nandy S, Kushwaha SPS, Dadhwal VK (2011) Forest degradation assessment in the upper catchment of the river Tons using remote sensing and GIS. Ecol Ind 11(2):509–513
Poffenberger M (1990) Joint management of forest lands: experiences from South Asia. A Ford Foundation Program statement
Poffenberger M, McGean B (1996) Village voices, forest choices. Joint Forest Management in India Oxford University Press Delhi
Rajan SR (2006) Modernizing nature: forestry and imperial eco-development 1800–1950. Oxford University Press
Rammer W, Seidl R (2015) Coupling human and natural systems: simulating adaptive management agents in dynamically changing forest landscapes. Glob Environ Chang 35:475–485
Rana E (2016) REDD+ and ecosystem services trade-offs and synergies in community forests of central Himalaya, Nepal
Rangarajan M (1999) Fencing the forest: conservation and ecological change in India's central provinces 1860–1914. pp xii+-245
Roy SB (1992) Forest protection committees in West Bengal. Econ Polit Wkly 1528–1530
Roy PS, Tomar S (2001) Landscape cover dynamics pattern in Meghalaya. Int J Remote Sens 22(18):3813–3825
Sarin M, Springate-Baginski O (2010) India's forest rights act: the anatomy of a necessary but not sufficient institutional reform. Research Programme Consortium for Improving Institutions for Pro-Poor Growth, Manchester
Saxena NC (1997) The saga of participatory forest management in India. CIFOR
Schlich W (1907) Schlich's manual of forestry: forest protection, by WR Fisher, being an English adaptation of "Der Forstschutz," by Richard Hess. 2d ed. 1907, vol 4. Bradbury, Agnew
Sen A (2021) A political ecology of forest conservation in India: communities, wildlife and the state. Routledge
Sen UK, Bhakat RK (2021) Conservation of resources by religious and social prohibitions by Santal communities in South West Bengal, India. Time Mind 14(1):3–32
Sen A, Lalhrietpui E (2006) Scheduled tribes (recognition of forest rights) bill: a view from anthropology and call for dialogue. Econ Polit Wkly 4205–4210
Sharma M, Kumar V, Kumar S (2024) A systematic review of urban sprawl and land use/land cover change studies in India. Sustain Environ 10(1):2331269
Sivaramakrishnan K (1999) Modern forests: statemaking and environmental change in colonial Eastern India. Stanford University Press
Stebbing EP (1925) Recent progress in Indian forestry. J Roy Soc Arts 74(3812):68–95

Sudhakar Reddy C, Jha CS, Dadhwal VK, Hari Krishna P, Vazeed Pasha S, Satish KV et al (2016) Quantification and monitoring of deforestation in India over eight decades (1930–2013). Biodivers Conserv 25(1):93–116
Sundar N (2000) Unpacking the 'joint' in joint forest management. Dev Chang 31(1):255–279
Sundar N (2001) Branching out: joint forest management in India. Oxford University Press
Thakur S, Maity D, Mondal I, Basumatary G, Ghosh PB, Das P, De TK (2021) Assessment of changes in land use, land cover, and land surface temperature in the mangrove forest of Sundarbans, northeast coast of India. Environ Dev Sustain 23(2):1917–1943

Chapter 3
Forest Fragmentation

Abstract Forest fragmentation represents a critical ecological challenge in tropical landscapes, particularly in densely populated regions where anthropogenic pressures intersect with biodiversity conservation imperatives. This study examines forest fragmentation dynamics in Jhargram District, West Bengal, over a three-decade period (1995–2025) using satellite-based remote sensing techniques and GIS analysis. Landsat imagery was processed to derive Normalized Difference Vegetation Index (NDVI) values, which were subsequently analyzed using the Landscape Fragmentation Tool (LFT) version 2.0 to quantify spatial patterns of habitat division and ecological connectivity. The analysis reveals alarming rates of forest degradation, with total forest area declining from 903.76 to 630.39 km^2, representing a 30.2% reduction over the study period. Most critically, large core forest areas (> 500 ha) experienced a 34.9% decline, decreasing from 416.78 to 271.39 km^2. Temporal analysis indicates that fragmentation processes accelerated dramatically between 2005 and 2015, when core areas declined from 355.4 to 184.89 km^2, coinciding with rapid infrastructure development and economic expansion. Model validation demonstrated exceptional performance with a Kappa coefficient of 0.83, coefficient of determination (R^2) of 0.87, and ROC Area Under Curve (AUC) of 0.842, confirming robust predictive capability. The fragmentation patterns exhibit complex dynamics, with patch forests showing variable trends while edge and perforated habitats demonstrated declining trajectories. The study's findings highlight the urgent need for integrated landscape management approaches that balance conservation objectives with socio-economic development requirements. The validated fragmentation model provides a valuable tool for evidence-based conservation planning and sustainable land-use policy development in tropical forest ecosystems.

Keywords Forest fragmentation · Normalized difference vegetation index · Landscape fragmentation tool · Jhargram · West Bengal

P. K. Shit and M. Hasanuzzaman, *Forest Landscape Dynamics in West Bengal*, Advances in Asian Human-Environmental Research,
https://doi.org/10.1007/978-3-032-23700-2_3

3.1 Introduction

Forest fragmentation represents a multi-dimensional ecological process whereby contiguous forest landscapes are progressively divided into smaller, more isolated patches, fundamentally altering ecosystem structure and function across multiple spatial and temporal scales (Matte et al. 2015). Conceptually distinct from outright deforestation, fragmentation involves four interrelated spatial processes: reduction in habitat amount, increasing isolation of patches, decreasing patch size, and increasing edge effects each with unique implications for biodiversity and ecosystem integrity (Fahrig 2003; Haddad et al. 2015). Quantitative analyses across global biomes reveal alarming trends; approximately 70% of remaining forest worldwide now exists within 1 km of a forest edge, and the mean distance to forest edge globally has decreased from 1975 m in 2000 to 1456 m by 2020, with tropical forests experiencing the most rapid fragmentation at rates exceeding 9.3% annually in some biodiversity hotspots (Haddad et al. 2015; Taubert et al. 2018; Betts et al. 2017). The drivers of fragmentation operate hierarchically and synergistically across scales, with broad-scale factors like macroeconomic policies and international market demands interacting with regional land tenure systems and local livelihood needs to create complex fragmentation trajectories that often exhibit non-linear dynamics and spatial contagion patterns (Meyfroidt et al. 2014; Rudel et al. 2009). Ecologically, fragmentation creates profound biophysical edge effects that typically extend 100–300 m into forest patches, altering microclimate conditions including increased temperature extremes (2–5 °C higher maximum temperatures), reduced humidity (15–30% lower at edges), greater wind disturbance, and elevated photosynthetically active radiation creating environmental filters that drive compositional shifts in biological communities (Harper et al. 2005; Magnago et al. 2014). These abiotic changes cascade through trophic networks; meta-analyses indicate that fragmentation reduces species richness by 13–75% and decreases biomass by 10–80% across taxonomic groups, with particularly severe impacts on large-bodied vertebrates, habitat specialists, and organisms with limited dispersal capabilities (Fletcher et al. 2018; Pfeifer et al. 2017). The genetic consequences of fragmentation further compound these impacts through processes including reduced gene flow, genetic drift, inbreeding depression, and disrupted metapopulation dynamics, with empirical evidence demonstrating 35–70% reductions in genetic diversity within small fragments compared to continuous habitat (Keyghobadi 2007; Schlaepfer et al. 2018). Beyond species-level impacts, fragmentation disrupts ecological processes including pollination (with 27% average reduction in fruit set documented in fragmented landscapes), seed dispersal (particularly affecting large-seeded plant species dependent on now-absent vertebrate dispersers), nutrient cycling (with significant carbon storage reductions of 10–25% in edge-affected areas), and trophic cascades as top predator extirpations in small fragments release mesopredator populations (Hagen et al. 2012; Magrach et al. 2014; Laurance et al. 2011). Critically, fragmentation creates synergistic interactions with other anthropogenic pressures; fragmented forests demonstrate 13–55% greater vulnerability to invasion by exotic species, increased susceptibility to fire propagation with edge areas 5–10 times more likely to burn than interior forest, and amplified climate change impacts due to microclimatic extremes that exceed regional warming projections (Didham et al. 2007; Briant et al. 2010; Alencar et al. 2015). The temporal dynamics of fragmentation introduce further complexity, with research documenting extinction debts wherein species persist temporarily in fragments before experiencing delayed extirpations, along with time-lagged ecosystem responses that may manifest decades after fragmentation events (Hylander and Ehrlén 2013; Metzger et al. 2009). Conservation responses to fragmentation have evolved from simple patch-focused approaches to sophisticated landscape-scale strategies emphasizing structural and functional connectivity through corridor establishment, buffer zone management, and matrix quality improvement, though empirical evaluations demonstrate variable effectiveness contingent upon landscape context and target species traits (Lindenmayer et al. 2008; Newmark et al. 2017). Recent conceptual advances emphasize the importance of habitat amount versus fragmentation per se, with the "habitat amount hypothesis" suggesting that patch size and isolation effects may be subordinate to total habitat availability within landscapes, though this remains contested particularly for area-sensitive species with limited dispersal abilities (Fahrig 2013; Hanski 2015; Fletcher et al. 2018). Contemporary research increasingly employs network theory, circuit theory, and agent-based models to quantify functional connectivity beyond structural measures, revealing complexity in how organisms perceive and navigate fragmented landscapes based on behavioral responses and matrix permeability (McRae et al. 2008; Zeller et al. 2012; Kool et al. 2013). From a policy perspective, fragmentation presents unique challenges because its incremental nature often escapes regulatory thresholds designed for outright deforestation, necessitating innovative governance approaches including cumulative effects assessment, cross-boundary coordination, and proactive land-use planning that maintains large intact forest blocks while strategically positioning human infrastructure to minimize fragmentation impacts (Lindenmayer and Fischer 2013; Laurance et al. 2014).

3.2 Forest Fragmentation of West Bengal

Forest fragmentation in West Bengal presents a complex spatial and ecological phenomenon characterized by the progressive division of once-contiguous forest landscapes into smaller, isolated patches with profound implications for biodiversity and ecosystem functioning (Mallick and Chakraborty 2018). Quantitative analyses using landscape metrics reveal that forest fragmentation has accelerated significantly since the 1970s, with patch density increasing by 236% between 1975 and 2018 while the mean patch size decreased from 328 ha to merely 67 ha across the state (Mandal and Chatterjee 2023). This fragmentation exhibits distinct regional patterns; the southwestern lateritic regions of Purulia, Bankura, and West Midnapore Districts experienced severe fragmentation through perforation processes, where small clearings within forest interiors gradually expanded and coalesced, primarily driven by fuelwood extraction and small-scale agricultural conversion (Sudhakar Reddy et al. 2016; Devi et al. 2013). In contrast, the Himalayan forests of Darjeeling and Kalimpong Districts have undergone edge-dominated fragmentation, with core forest area declining by approximately 41% between 1990 and 2020 due to linear infrastructure development, particularly road networks that increased from 0.8 to 2.3 km/km^2 during this period, creating barrier effects for wildlife movement (Hussain et al. 2024; Mandal and Chatterjee 2023). The fragmentation process in the Sundarbans mangrove ecosystem demonstrates yet another pattern, with isolation of forest patches occurring through hydrological modifications and embankment construction that disrupted natural tidal regeneration processes (Mukhopadhyay et al. 2021). Multi-scale analysis indicates that fragmentation operates hierarchically, with different drivers predominating at different spatial scales, while agricultural expansion remains the primary driver at the landscape scale, accounting for 62% of all fragmentation events, at finer scales infrastructure development (27%), settlement expansion (18%), and plantation forestry (14%) emerge as significant contributors (Mitra et al. 2019; Chakraborty et al. 2017). Temporal analysis using time-series metrics reveals that fragmentation has not proceeded linearly but rather exhibits threshold effects, with particularly rapid transitions occurring during 1985–1995 and 2005–2015, periods corresponding to significant policy shifts and land market transformations (Hussain et al. 2024; Mallick 2016). The ecological consequences of this fragmentation are profound and taxonomically variable; studies document a 47% reduction in mammalian diversity in forest patches smaller than 50 ha compared to larger contiguous tracts, with area-sensitive species like the Indian elephant (*Elephas maximus*) and gaur (*Bos gaurus*) experiencing range contractions exceeding 60% over three decades (Garmendia et al. 2013; Ramesh et al. 2012). Avian communities demonstrate varying resilience to fragmentation, with riverine and edge-preferring species increasing while interior forest specialists declined by approximately 38% in richness (Alfred et al. 2012; Penjor et al. 2022). Plant communities exhibit cascading responses to fragmentation with altered regeneration dynamics; edge-affected forests show 215% higher abundance of pioneer species but 73% reduction in late-successional taxa, alongside disruptions to pollination networks particularly affecting dioecious species that showed 29% lower fruit set in fragmented habitats (Ismail et al. 2014; Arroyo-Rodríguez et al. 2017). Beyond biodiversity impacts, fragmentation has amplified human-wildlife conflicts, particularly elephant depredation which increased by 114% in areas with fragmentation-induced migration corridor disruptions (Roy and Sukumar 2017). Modeling approaches incorporating structural and functional connectivity metrics suggest that fragmentation has created complex metapopulation dynamics for many species, with certain fragmented landscapes maintaining connectivity through stepping-stone patches while others have crossed critical thresholds beyond which demographic rescue effects become highly improbable (Dutta et al. 2016). Recent conservation approaches have begun addressing fragmentation through corridor restoration initiatives, particularly in the "Elephant Corridor Project" connecting Dalma and Mayurjharna Elephant Reserves, though implementation challenges persist due to competing land uses and complex tenure arrangements (Dhyani et al. 2020; Kumari et al. 2024).

3.3 Forest Fragmentation Assessment: GIS Tools and Techniques

Forest fragmentation analysis has evolved significantly with the advancement of Geographic Information Systems (GIS), offering researchers and practitioners powerful tools to quantify landscape changes. Modern GIS platforms provide specialized modules and extensions dedicated to fragmentation metrics calculation, enabling comprehensive assessment of forest ecosystem integrity. FRAGSTATS, developed by McGarigal (1995), remains one of the most widely implemented software packages, calculating over 50 landscape metrics including patch density, edge density, and core area indices. The Patch Analyst extension for ArcGIS (Rempel et al. 2012) offers similar functionality with improved integration into mainstream GIS workflows. More recently, machine learning approaches have been incorporated into fragmentation analysis through tools like LandscapeR (Stupariu et al. 2022), which employs random forest algorithms to detect subtle fragmentation patterns across temporal scales. For continuous landscape assessment, moving-window techniques have gained prominence, with spatial resolution considerations being critical in determining appropriate window sizes (Hagen-Zanker 2016). These technologies have transformed forest fragmentation from a qualitative observation to a quantifiable ecological parameter, enabling evidence-based conservation planning and restoration prioritization. Current methodological advancements focus on integrating satellite imagery analysis with fragmentation metrics through cloud computing platforms like Google Earth Engine, allowing for near-real-time monitoring of forest connectivity at regional scales (Yang et al. 2022).

3.4 A Case Study on Forest Fragmentation in the Jhargram District of West Bengal

The Jhargram District of West Bengal presents a compelling case study for understanding forest fragmentation dynamics in the eastern Indian landscape, where anthropogenic pressures intersect with ecological conservation imperatives. Located in the southwestern region of West Bengal, Jhargram District encompasses approximately 3037 km^2 and was carved out as a separate administrative unit from West Midnapore District in 2017. The region's forest ecosystem, primarily characterized by tropical dry deciduous forests interspersed with sal (*Shorea robusta*)-dominated patches, has experienced significant fragmentation over the past four decades due to a complex interplay of socio-economic and developmental factors (Sahoo et al. 2023).

Historical analysis reveals that Jhargram's forest cover, which originally extended across nearly 60% of the district's geographical area in the 1970s, has been progressively fragmented through multiple anthropogenic drivers (Mandal and Chatterjee 2021). Agricultural expansion, particularly the conversion of forest margins to paddy cultivation and horticultural plantations, has created a mosaic landscape where forest patches are increasingly isolated by agricultural matrices (Castella et al. 2013). The construction of transportation infrastructure, including the expansion of National Highway and the development of rural connectivity projects, has further contributed to habitat fragmentation by creating linear barriers that impede wildlife movement and gene flow between forest patches (Theobald et al. 2011).

The fragmentation pattern in Jhargram exhibits characteristics typical of edge-dominated landscapes, where the ratio of forest edge-to-interior habitat has increased substantially. Satellite imagery analysis conducted between 1990 and 2020 demonstrates that while the total forest cover declined by approximately 18%, the number of forest fragments increased by 340%, with the average patch size decreasing from 847 to 312 ha (Sahoo et al. 2023). This fragmentation has resulted in the creation of numerous small, isolated forest patches that are particularly vulnerable to edge effects, including altered microclimate conditions, increased susceptibility to invasive species colonization, and reduced habitat quality for forest-dependent fauna.

The ecological consequences of forest fragmentation in Jhargram are manifested through multiple pathways that threaten biodiversity conservation and ecosystem functioning (Mandal and Chatterjee 2021). Large mammalian species, including Asian elephants (*Elephas maximus*), have experienced habitat compression, leading to increased human-elephant conflict incidents, particularly during crop ripening seasons when elephants venture into agricultural areas (Shaffer et al. 2019). The fragmentation has also affected the population dynamics of arboreal species such as the Indian giant squirrel (*Ratufa indica*) and various primate species, whose dispersal capabilities are limited by the agricultural matrix separating forest patches. Furthermore, the disruption of pollination networks due to habitat fragmentation has implications for plant reproductive success and forest regeneration processes, as documented in studies examining the breeding systems of native tree species in fragmented landscapes (Breed et al. 2015).

The socio-economic dimensions of forest fragmentation in Jhargram are equally complex, reflecting the livelihood dependencies of local communities on forest resources. Approximately 35% of the district's population belongs to scheduled tribes, including the Santhal, Munda, and Kol communities, who have traditionally maintained subsistence relationships with forest ecosystems through practices such as minor forest produce collection, traditional agriculture, and livestock grazing (Sahoo et al. 2023). The fragmentation process has disrupted these traditional resource access patterns, creating conflicts between conservation objectives and livelihood securities. Additionally, the fragmented landscape has altered the provision of ecosystem services, including watershed regulation, carbon sequestration, and climate moderation, with implications for regional environmental stability and agricultural productivity (Qiu et al. 2020).

Contemporary conservation efforts in Jhargram District have attempted to address fragmentation through various landscape-level interventions, including the establishment of biological corridors, community-based forest management initiatives, and agroforestry programs designed to enhance connectivity between isolated forest patches. The Joint Forest Management (JFM) program, implemented in collaboration with Forest Protection Committees, has shown promising results in selected areas where community participation has been effectively mobilized for forest restoration activities (Tiwary 2019). However, the success of these interventions remains spatially variable and dependent on local socio-economic conditions, institutional capacity, and the degree of community engagement in conservation activities.

The Jhargram case study illuminates the broader challenges of managing forest fragmentation in densely populated tropical landscapes where conservation goals must be balanced against developmental aspirations and livelihood requirements. The district's experience demonstrates that effective fragmentation mitigation requires integrated approaches that address both ecological and socio-economic dimensions of landscape change, emphasizing the importance of participatory planning processes that incorporate traditional ecological knowledge and community-based management systems. Future research directions should focus on developing spatially explicit models that can predict fragmentation trajectories under different development scenarios, thereby informing evidence-based conservation planning and sustainable landscape management strategies.

3.4.1 Materials and Methods

Dataset

We utilized Landsat satellite imagery from the years 1995, 2005, 2015, and 2025, all of which were downloaded from the USGS website. These images were used to extract the Normalized Difference Vegetation Index (NDVI) in order to distinguish between forested and non-forested areas. To minimize temporal inconsistencies, all imagery was selected as close as possible to the same date across the different years. Preprocessing of images, including various radiometric and geometric corrections, was carried out using QGIS software, following the methodology outlined by Kganyago et al. (2020).

NDVI

The Normalized Difference Vegetation Index (NDVI) is a widely recognized remote sensing metric used to assess vegetation health and monitor terrestrial productivity. Developed by Pettorelli (2013), NDVI quantifies photosynthetic activity by measuring the contrast between near-infrared (NIR) and red spectral reflectance. NDVI values range from -1 to $+1$, where lower values (closer to -1) indicate sparse or stressed vegetation, such as barren land or water bodies, while higher values (approaching $+1$) correspond to dense, healthy vegetation like forests or thriving crops (Bhandari et al. 2012). The index is calculated using the formula:

$$\text{NDVI} = (\text{NIR} - \text{RED})/(\text{NIR} + \text{RED}) \tag{3.1}$$

Forest Fragmentation Model

The Forest Fragmentation Model serves as a widely utilized analytical framework for quantifying deforestation patterns and assessing spatial discontinuities in forest cover (Sharma et al. 2016). In this study, forest fragmentation was analyzed using the Landscape Fragmentation Tool (LFT) version 2.0, an ArcGIS extension originally developed by the Center for Land Use Education and Research (CLEAR) in 2002. This tool, validated in prior studies (Randhawa et al. 2024), classifies fragmented landscapes into distinct structural categories based on established methodologies (Vogt et al. 2007). Input data for the model included binary forest/non-forest maps derived from NDVI-based classification. The output delineated six fragmentation classes: **patch, edge, perforated, low core** (< 250 acres)**, medium core** (250–500 acres), and **high core** (> 500 acres). A 25-m edge buffer was applied to differentiate transitional zones between forested and non-forested areas, consistent with best practices for edge-effect quantification in fragmented ecosystems.

Method of Validation

Model validation is a critical step in ensuring the reliability of scientific research (Hasanuzzaman et al. 2022). In this study, the accuracy of the 2025 forest fragmentation model was assessed using 153 ground-truth points collected during field surveys. The performance of the fragmentation analysis was evaluated through a Receiver Operating Characteristic (ROC) curve and four statistical metrics: Mean Absolute Error (MAE), Root Mean Square Error (RMSE), Kappa Index (K), and Coefficient of Determination (R^2). These metrics were selected to quantify prediction errors, classification agreement, and overall model robustness. The equations applied in this validation process are as follows:

$$\text{RMSE} = \sqrt{\frac{1}{n}\sum_{i=1}^{i=n}(X_{\text{ei}} - X_{\text{oi}})^2} \tag{3.2}$$

$$\text{MAE} = \frac{1}{n}\sum_{i=1}^{i=n}|X_{\text{ei}} - X_{\text{oi}}| \tag{3.3}$$

$$\text{Kappa}(k) = \frac{P_c - P_{\text{cxp}}}{1 - P_{\text{cxp}}} \tag{3.4}$$

where P_c indicates numeral of pixels to be matched accurately as forest or non-forest pixels; P_{cxp} denotes estimated results; X_{oi} and X_{ei} are the ith observed and model predicted values, respectively; and n is the amount of data point (Khosravi et al. 2019).

This study incorporates Receiver Operating Characteristic (ROC) curves to evaluate the predictive accuracy of the forest fragmentation models. The ROC curve graphically assesses model performance by plotting sensitivity (true positive rate) on the Y-axis against 1-specificity (false positive rate) on the X-axis, distinguishing between forested and non-forested areas (Hasanuzzaman et al. 2022). The Area Under the Curve (AUC), calculated from the ROC plot, quantifies overall model efficacy, with values ranging from 0 to 1. An AUC closer to 1 indicates high predictive accuracy, whereas values near 0.5 suggest performance no better than random chance (Karimi-Rizvandi et al. 2021). In this analysis, AUC values exceeding 0.8 were interpreted as robust evidence of model reliability, aligning with established thresholds for ecological forecasting.

3.4.2 *Results and Discussion*

Forest Fragmentation Dynamics and Temporal Patterns

The analysis of forest fragmentation in Jhargram District over the three-decade period (1995–2025) reveals significant spatiotemporal changes in forest landscape structure, with implications for biodiversity conservation and ecosystem integrity.

The Normalized Difference Vegetation Index (NDVI) analysis demonstrates a progressive decline in vegetation density across the study period, indicating substantial forest cover loss and degradation processes operating at the landscape level (Fig. 3.1).

The forest fragmentation assessment reveals a concerning trajectory of habitat loss and structural simplification. Total forest area decreased from 903.76 km^2 in 1995 to 630.39 km^2 in 2025, representing a net loss of 273.37 km^2 or approximately 30.2% of the original forest cover over the 30-year study period (Table 3.1). This rate of forest loss (approximately 1% per year) exceeds the global average for tropical dry forests and reflects the intense anthropogenic pressures operating in this densely populated region of eastern India. The fragmentation pattern exhibits distinct characteristics across different forest categories. Core forest areas larger than 500 ha, which represent the most ecologically valuable habitats for forest-dependent species, experienced the most dramatic decline from 416.78 km^2 in 1995 to 271.39 km^2 in 2025, constituting a 34.9% reduction (Figs. 3.2, 3.3, 3.4 and 3.5).

This substantial loss of large core areas is particularly concerning as these habitats support viable populations of wide-ranging species and maintain essential ecological processes such as seed dispersal networks and predator–prey dynamics. The temporal analysis reveals that the most severe losses occurred between 2005 and 2015, when large core areas declined from 355.4 to 184.89 km^2, suggesting an acceleration of fragmentation processes during this decade (Figs. 3.3 and 3.4).

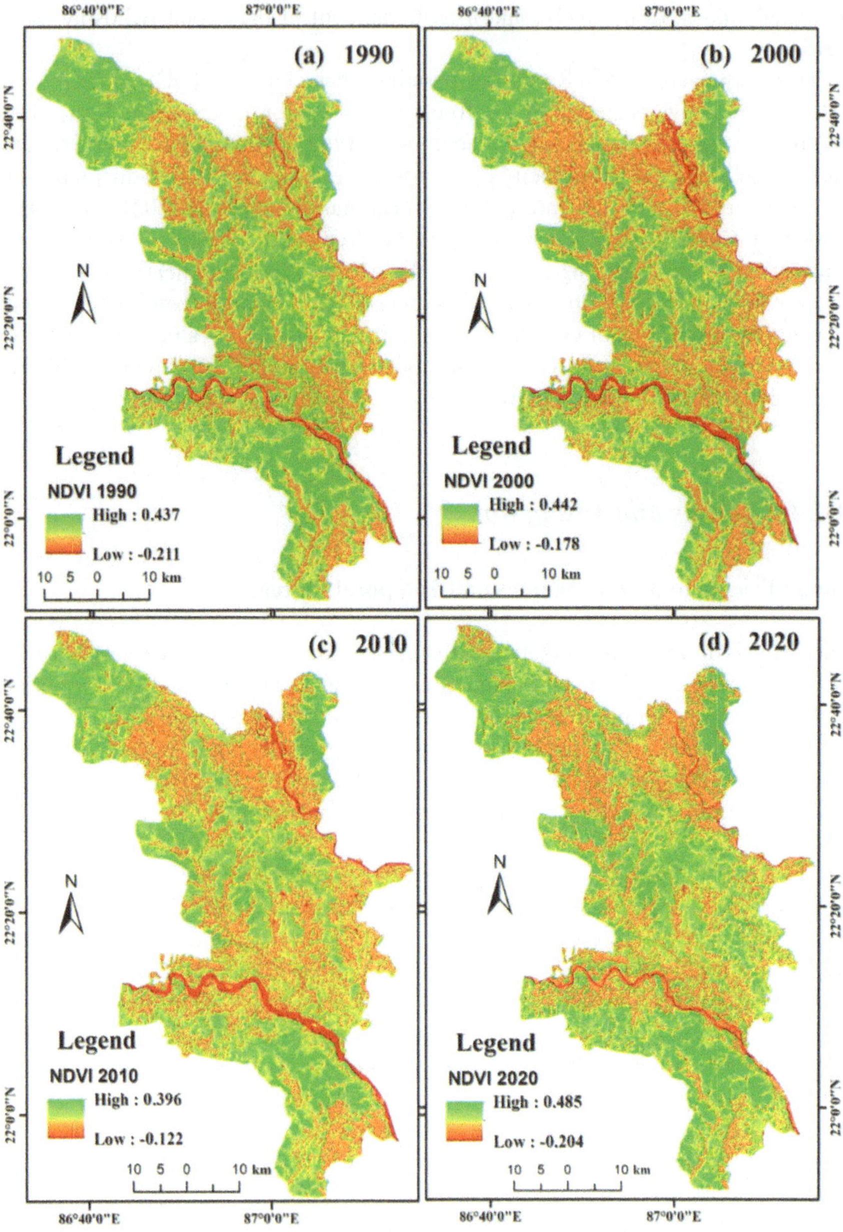

Fig. 3.1 Normalized difference vegetation index (NDVI) of Jhargram District: **a** 1995, **b** 2005, **c** 2015, and **d** 2025

Table 3.1 Forest fragmentation changes from 1995 to 2025

Forest fragmentation	1995	2005	2015	2025
Patch	73.74	55.39	93.26	63.66
Edge	228.14	174.53	217.55	153.27
Perforated	48.5	34.13	45.54	39.97
Core (< 250)	104.83	86.51	111.59	74.2
Core (250–500)	31.77	30.86	23.82	27.9
Core (> 500)	416.78	355.4	184.89	271.39
Total	903.76	736.82	676.65	630.39

Conversely, patch forests, representing the most fragmented and isolated habitat remnants, showed variable trends with increases during certain periods (reaching 93.26 km^2 in 2015) before declining to 63.66 km^2 in 2025 (Fig. 3.5). This fluctuation likely reflects the dynamic nature of small forest remnants, which are subject to rapid conversion, natural regeneration, and edge effects that can cause rapid appearance and disappearance of small patches.

The increase in patch forests during the 2005–2015 period coincides with the dramatic reduction in large core areas, suggesting that fragmentation processes were actively breaking down larger forest blocks into smaller, isolated units. Edge habitats, characterized by transitional zones between forest and non-forest areas, demonstrated a consistent declining trend from 228.14 km^2 in 1995 to 153.27 km^2 in 2025 (Fig. 3.6).

This reduction in edge habitat area, while potentially positive from a forest interior perspective, likely reflects the overall forest loss rather than improved habitat quality, as the edge-to-interior ratio may have actually increased due to the creation of more numerous but smaller forest patches.

Perforated forests, indicating areas with internal gaps and openings, showed relatively stable patterns with fluctuations around 40–48 km^2 throughout the study period. The persistence of perforated forest patterns suggests ongoing internal disturbance processes, including selective logging, grazing, and local resource extraction activities, that create openings within otherwise continuous forest cover (Fig. 3.7).

Model Performance and Validation

The forest fragmentation model demonstrated exceptional predictive accuracy, as evidenced by multiple validation metrics (Fig. 3.8). The Kappa coefficient of 0.83 indicates strong agreement between predicted and observed fragmentation patterns, substantially exceeding the 0.70 threshold typically considered acceptable in ecological modeling studies. This high agreement suggests that the model successfully captures the underlying drivers and processes governing forest fragmentation in the study area. The coefficient of determination (R^2) value of 0.87 demonstrates that the model explains 87% of the variance in forest fragmentation patterns, indicating

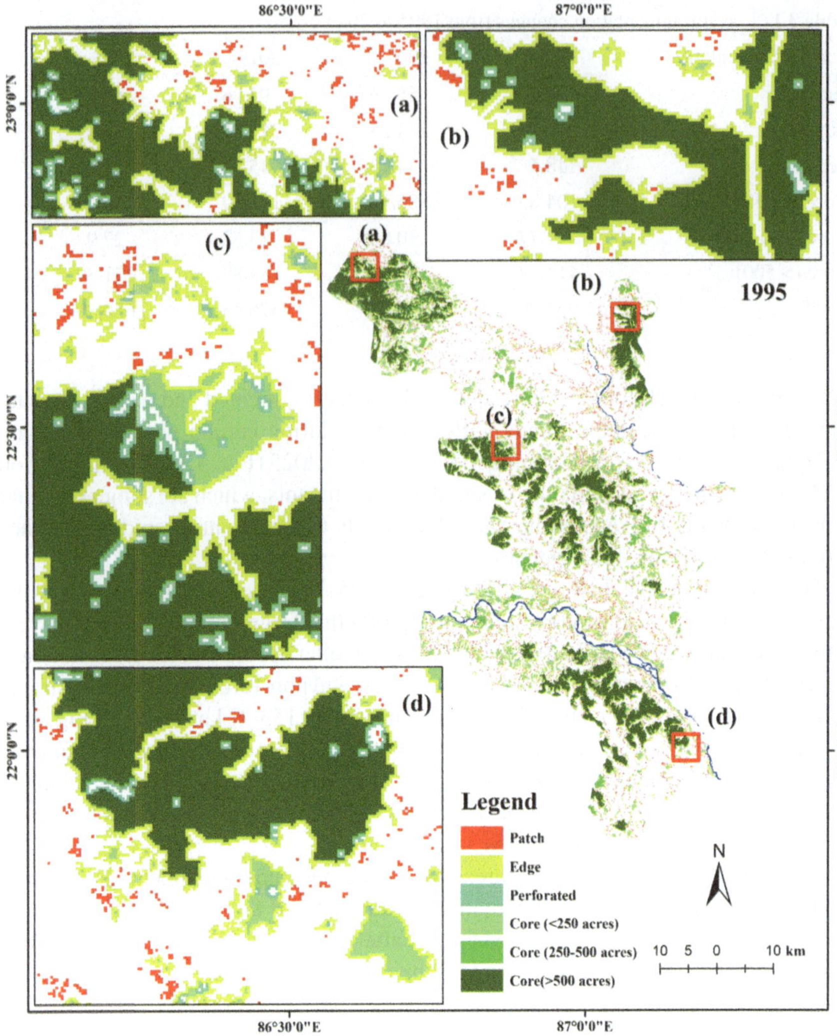

Fig. 3.2 Forest fragmentation map of the study area of 1995

robust explanatory power and suggesting that the selected variables and modeling approach effectively represent the complex interactions driving landscape change.

The low error rates, with Mean Absolute Error (MAE) of 0.29 and Root Mean Square Error (RMSE) of 27, further confirm the model's precision in predicting fragmentation outcomes. The Receiver Operating Characteristic (ROC) analysis yielded an Area Under the Curve (AUC) value of 0.842, surpassing the 0.8 threshold associated with high predictive reliability in ecological applications (Fig. 3.9). This performance metric indicates that the model demonstrates excellent discriminatory ability in distinguishing between different fragmentation categories and can reliably predict

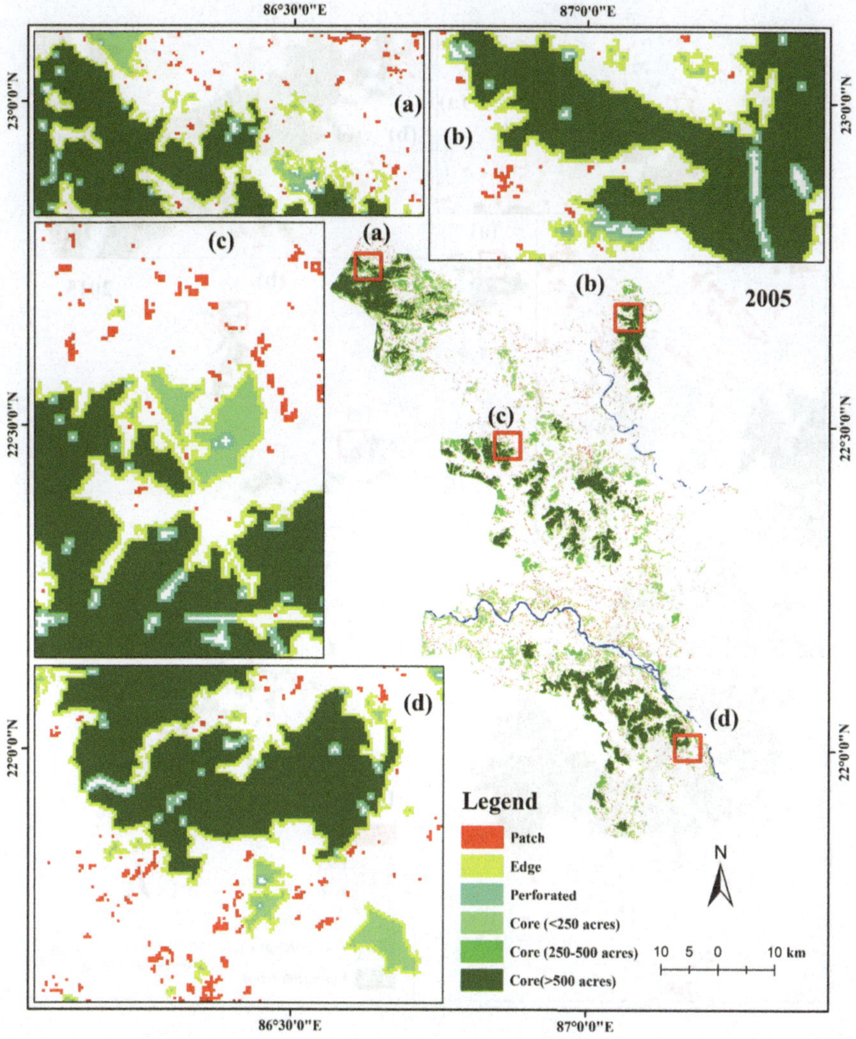

Fig. 3.3 Forest fragmentation map of the study area of 2005

future fragmentation scenarios under similar environmental and socio-economic conditions.

Ecological and Conservation Implications

The observed fragmentation patterns have profound implications for biodiversity conservation and ecosystem functioning in the Jhargram District. The substantial

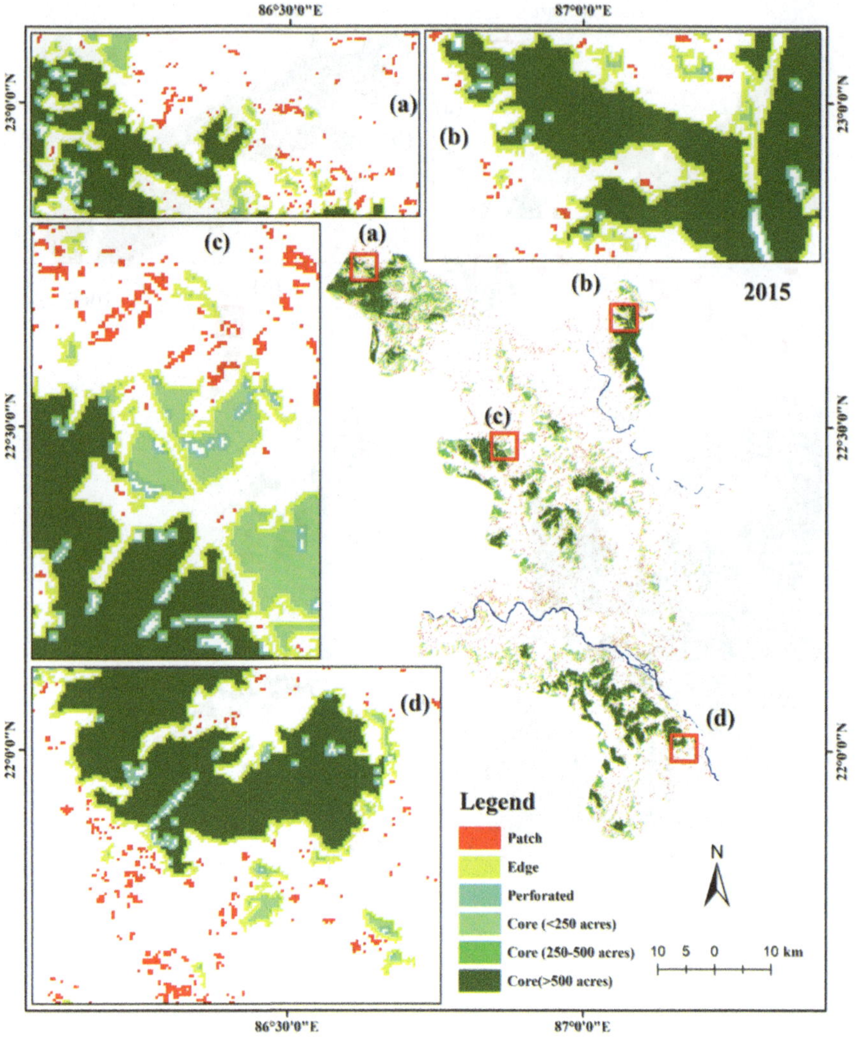

Fig. 3.4 Forest fragmentation map of the study area of 2015

loss of large core forest areas represents a critical threat to species requiring extensive, undisturbed habitats, including large mammals, forest-dependent birds, and specialized plant communities.

The fragmentation process has likely resulted in population isolation, reduced genetic diversity, and increased extinction risks for sensitive species. The creation of smaller, isolated forest patches increases edge effects, which can penetrate 100–300 m into forest interiors, altering microclimate conditions, increasing invasive species establishment, and disrupting natural ecological processes. Given the observed reduction in average patch size, a significant proportion of remaining forest

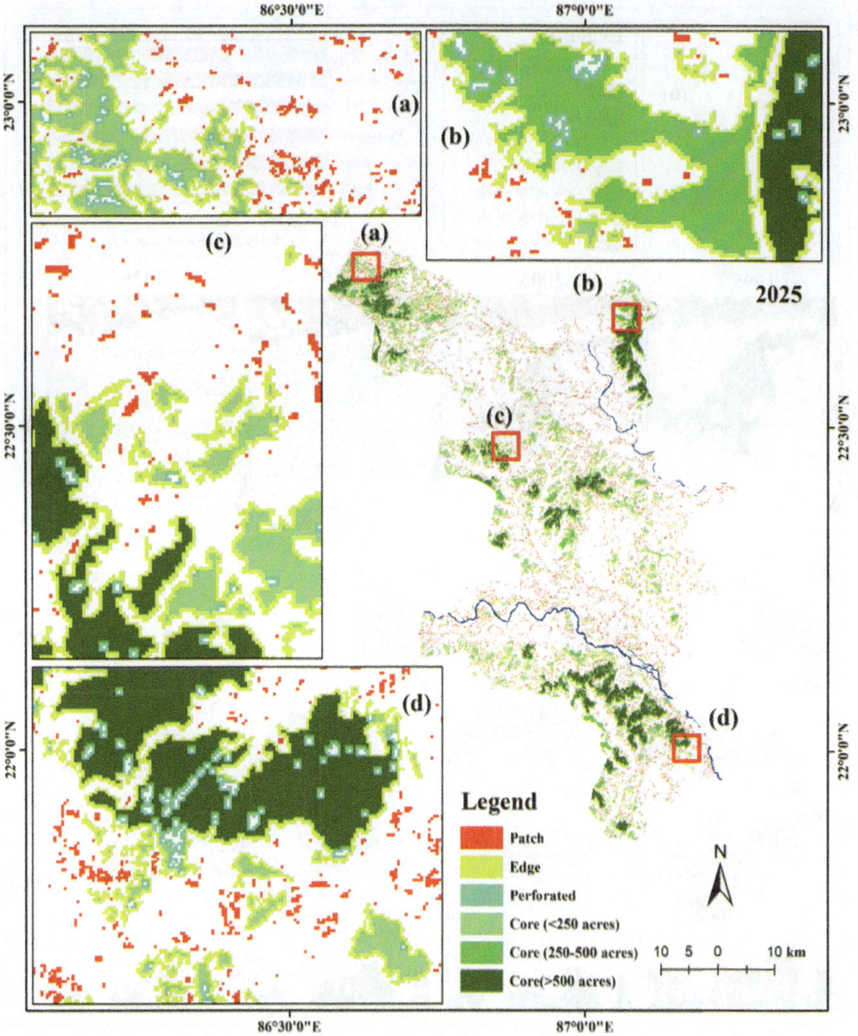

Fig. 3.5 Forest fragmentation map of the study area of 2025

area is now subject to edge influences, potentially compromising habitat quality even in apparently intact areas. The temporal acceleration of fragmentation between 2005 and 2015 coincides with rapid economic development and infrastructure expansion in West Bengal, suggesting that development pressures are the primary drivers of forest loss. The subsequent deceleration between 2015 and 2025 may reflect the implementation of stronger conservation policies and forest protection measures, though continued monitoring is essential to assess the long-term effectiveness of these interventions. The validated fragmentation model provides a valuable tool for conservation planning and landscape management in the region. The high predictive

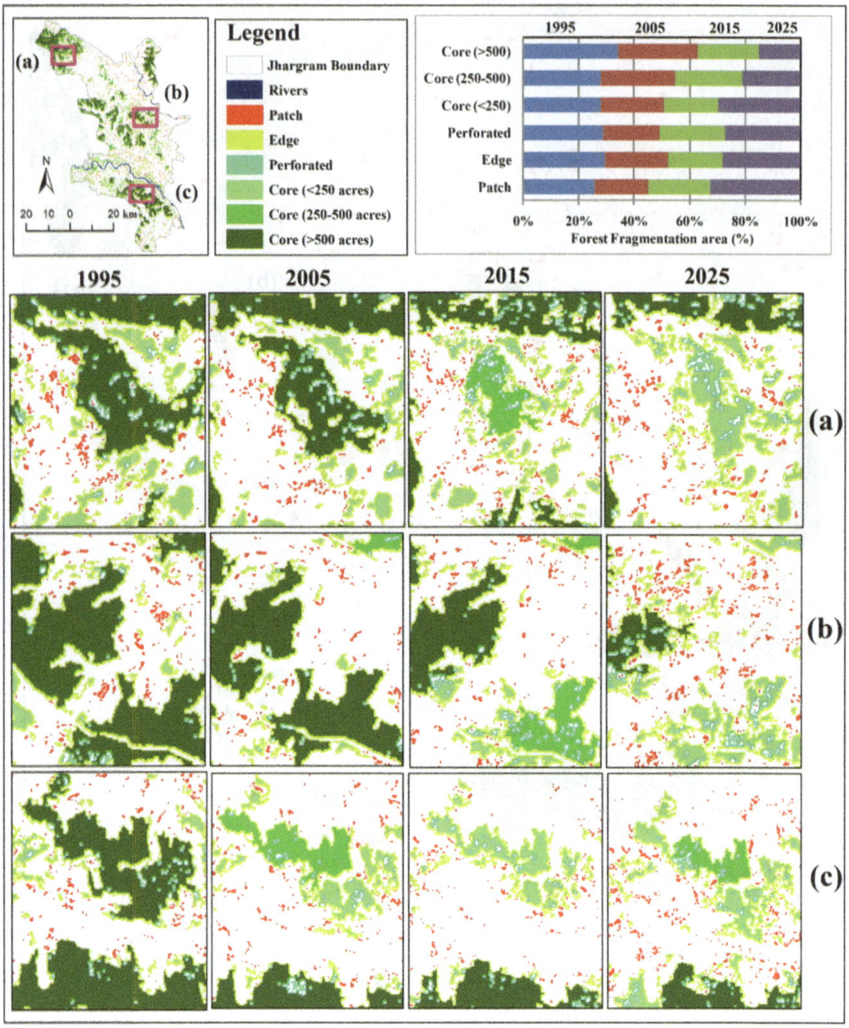

Fig. 3.6 Forest fragmentation change map from 1995 to 2025 of the Jhargram District, West Bengal

accuracy enables reliable forecasting of fragmentation trajectories under different development scenarios, supporting evidence-based decision-making for conservation prioritization and sustainable land-use planning initiatives.

Fig. 3.7 **a** A dry forest pathway cutting through a degraded sal forest, indicating reduced canopy cover and signs of anthropogenic disturbance. **b** Recently felled tree trunks lying on the forest floor, demonstrating active timber extraction and forest resource removal. **c** A patchy young plantation area with scattered saplings, reflecting ongoing afforestation or regeneration efforts in previously disturbed land. **d** An open forest tract with visible tree stumps and sparse vegetation, showing evidence of logging pressure and forest degradation. **e** An agricultural field with a pronounced soil embankment and erosion features, illustrating land-use conversion and associated soil degradation. **f** Stacked wooden poles and cut timber stored near a rural settlement, highlighting local dependence on forest biomass for construction and livelihood purposes

Fig. 3.8 **a** A dry deciduous Sal Forest with a sparse understorey and accumulated leaf litter, indicating seasonal moisture stress and low ground vegetation. **b** A straight forest track passing through a moderately dense stand, showing managed woodland conditions and human accessibility within the forest interior. **c** A regenerating forest patch with dense young saplings and fresh ground cover, reflecting natural regeneration or assisted plantation growth. **d** A plantation area with uniformly spaced tall trees and thick shrub growth beneath, illustrating monoculture management and understorey development. **e** A comparatively healthy sal stand with improved canopy continuity and rich green undergrowth, suggesting better moisture availability and forest recovery. **f** A degraded forest segment marked by thin, sparsely foliated tree trunks and dense invasive undergrowth, indicating ecological stress and possible past disturbance

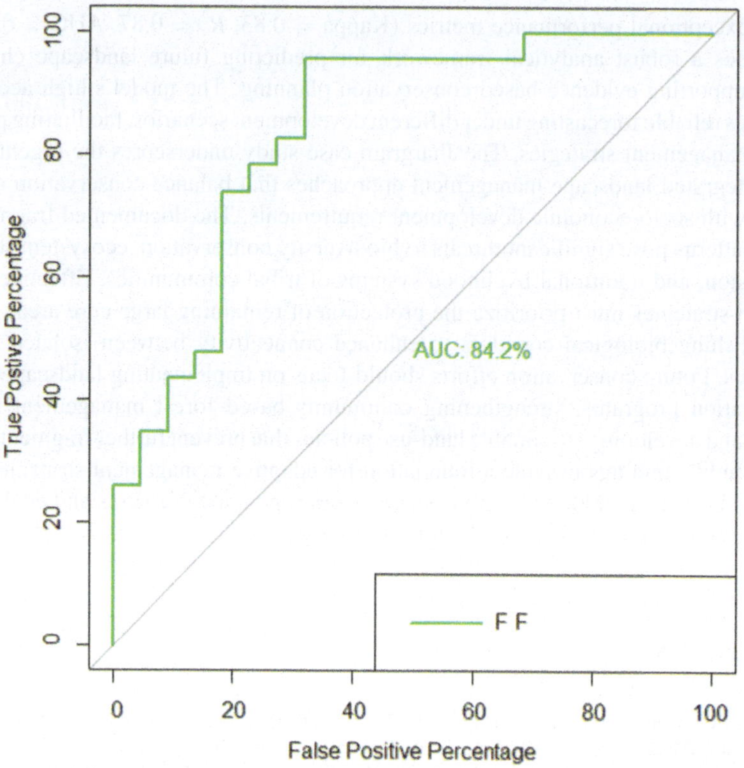

Fig. 3.9 ROC for model validation

3.4.3 Conclusion

The comprehensive analysis of forest fragmentation in Jhargram District over three decades (1995–2025) provides critical insights into the dynamics of landscape degradation in tropical dry deciduous forests of eastern India. The study reveals alarming rates of habitat loss, with total forest area declining by 30.2% (273.37 km^2) over the study period, significantly exceeding global deforestation rates for similar ecosystems. The most concerning finding is the 34.9% reduction in large core forest areas (> 500 ha), which represent the most ecologically valuable habitats for maintaining biodiversity and ecosystem functioning. The temporal analysis demonstrates that fragmentation processes accelerated dramatically between 2005 and 2015, coinciding with rapid infrastructure development and economic expansion in West Bengal. This period witnessed the conversion of large, contiguous forest blocks into smaller, isolated patches, fundamentally altering the landscape's ecological structure and connectivity. The subsequent deceleration between 2015 and 2025 suggests that conservation interventions may be moderating fragmentation rates, though continued vigilance is essential. The validated forest fragmentation model, with exceptional performance metrics (Kappa = 0.83, $R^2 = 0.87$, AUC = 0.842), provides a robust analytical framework for predicting future landscape changes and supporting evidence-based conservation planning. The model's high accuracy enables reliable forecasting under different development scenarios, facilitating proactive management strategies. The Jhargram case study underscores the urgent need for integrated landscape management approaches that balance conservation objectives with socio-economic development requirements. The documented fragmentation patterns pose significant threats to biodiversity conservation, ecosystem service provision, and traditional livelihood systems of tribal communities. Effective mitigation strategies must prioritize the protection of remaining large core areas while establishing biological corridors to enhance connectivity between isolated forest patches. Future conservation efforts should focus on implementing landscape-scale restoration programs, strengthening community-based forest management initiatives, and developing sustainable land-use policies that prevent further fragmentation. The study's findings provide a foundation for adaptive management strategies that can address the complex interplay between anthropogenic pressures and ecological conservation imperatives in tropical forest landscapes.

References

Alencar AA, Brando PM, Asner GP, Putz FE (2015) Landscape fragmentation, severe drought, and the new Amazon forest fire regime. Ecol Appl 25(6):1493–1505

Alfred JRB, Sanyal AK, Tiwari S, Mitra S (2012) Status of biodiversity of West Bengal. Final technical report. Department of Environment. Kolkata, West Bengal, India

Arroyo-Rodríguez V, Melo FP, Martínez-Ramos M, Bongers F, Chazdon RL, Meave JA et al (2017) Multiple successional pathways in human-modified tropical landscapes: new insights from forest succession, forest fragmentation and landscape ecology research. Biol Rev 92(1):326–340

Betts MG, Wolf C, Ripple WJ, Phalan B, Millers KA, Duarte A et al (2017) Global forest loss disproportionately erodes biodiversity in intact landscapes. Nature 547(7664):441–444

Bhandari AK, Kumar A, Singh GK (2012) Feature extraction using Normalized Difference Vegetation Index (NDVI): a case study of Jabalpur city. Procedia Technol 6:612–621

Breed MF, Ottewell KM, Gardner MG, Marklund MH, Dormontt EE, Lowe AJ (2015) Mating patterns and pollinator mobility are critical traits in forest fragmentation genetics. Heredity 115(2):108–114

Briant G, Gond V, Laurance SG (2010) Habitat fragmentation and the desiccation of forest canopies: a case study from eastern Amazonia. Biol Cons 143(11):2763–2769

Castella JC, Lestrelin G, Hett C, Bourgoin J, Fitriana YR, Heinimann A, Pfund JL (2013) Effects of landscape segregation on livelihood vulnerability: moving from extensive shifting cultivation to rotational agriculture and natural forests in northern Laos. Hum Ecol 41(1):63–76

Chakraborty A, Ghosh A, Sachdeva K, Joshi PK (2017) Characterizing fragmentation trends of the Himalayan forests in the Kumaon region of Uttarakhand, India. Eco Inform 38:95–109

Devi BS, Murthy MSR, Debnath B, Jha CS (2013) Forest patch connectivity diagnostics and prioritization using graph theory. Ecol Model 251:279–287

Dhyani S, Bartlett D, Kadaverugu R, Dasgupta R, Pujari P, Verma P (2020) Integrated climate sensitive restoration framework for transformative changes to sustainable land restoration. Restor Ecol 28(5):1026–1031

Didham RK, Tylianakis JM, Gemmell NJ, Rand TA, Ewers RM (2007) Interactive effects of habitat modification and species invasion on native species decline. Trends Ecol Evol 22(9):489–496

Dutta T, Sharma S, McRae BH, Roy PS, DeFries R (2016) Connecting the dots: mapping habitat connectivity for tigers in central India. Reg Environ Change 16(Suppl 1):53–67

Fahrig L (2003) Effects of habitat fragmentation on biodiversity. Annu Rev Ecol Evol Syst 34(1):487–515

Fahrig L (2013) Rethinking patch size and isolation effects: the habitat amount hypothesis. J Biogeogr 40(9):1649–1663

Fletcher RJ Jr, Didham RK, Banks-Leite C, Barlow J, Ewers RM, Rosindell J et al (2018) Is habitat fragmentation good for biodiversity? Biol Cons 226:9–15

Garmendia A, Arroyo-Rodríguez V, Estrada A, Naranjo EJ, Stoner KE (2013) Landscape and patch attributes impacting medium-and large-sized terrestrial mammals in a fragmented rain forest. J Trop Ecol 29(4):331–344

Haddad NM, Brudvig LA, Clobert J, Davies KF, Gonzalez A, Holt RD et al (2015) Habitat fragmentation and its lasting impact on Earth's ecosystems. Sci Adv 1(2):e1500052

Hagen M, Kissling WD, Rasmussen C, De Aguiar MA, Brown LE, Carstensen DW et al (2012) Biodiversity, species interactions and ecological networks in a fragmented world. In: Advances in ecological research, vol 46. Academic Press, pp 89–210

Hagen-Zanker A (2016) A computational framework for generalized moving windows and its application to landscape pattern analysis. Int J Appl Earth Obs Geoinf 44:205–216

Hanski I (2015) Habitat fragmentation and species richness. J Biogeogr 42(5):989–993

Harper KA, Macdonald SE, Burton PJ, Chen J, Brosofske KD, Saunders SC et al (2005) Edge influence on forest structure and composition in fragmented landscapes. Conserv Biol 19(3):768–782

Hasanuzzaman M, Islam A, Bera B, Shit PK (2022) A comparison of performance measures of three machine learning algorithms for flood susceptibility mapping of river Silabati (tropical river, India). Phys Chem Earth Parts a/b/c 127:103198

Hussain S, Sharma S, Bhatti RC, Singh AN (2024) Sustainability in the Indian Himalayan Region: understandings the ecosystem services, climate change impacts, land use shifts and their threats. In: The Himalayas in the Anthropocene: environment and development. Springer Nature Switzerland, Cham, pp 33–57

Hylander K, Ehrlén J (2013) The mechanisms causing extinction debts. Trends Ecol Evol 28(6):341–346

Ismail SA, Ghazoul J, Ravikanth G, Kushalappa CG, Uma Shaanker R, Kettle CJ (2014) Fragmentation genetics of Vateria indica: implications for management of forest genetic resources of an endemic dipterocarp. Conserv Genet 15(3):533–545

Karimi-Rizvandi S, Goodarzi HV, Afkoueieh JH, Chung IM, Kisi O, Kim S, Linh NTT (2021) Groundwater-potential mapping using a self-learning bayesian network model: a comparison among metaheuristic algorithms. Water 13(5):658

Keyghobadi N (2007) The genetic implications of habitat fragmentation for animals. Can J Zool 85(10):1049–1064

Kganyago M, Mhangara P, Alexandridis T, Laneve G, Ovakoglou G, Mashiyi N (2020) Validation of sentinel-2 leaf area index (LAI) product derived from SNAP toolbox and its comparison with global LAI products in an African semi-arid agricultural landscape. Remote Sens Lett 11(10):883–892

Khosravi K, Shahabi H, Pham BT, Adamowski J, Shirzadi A, Pradhan B et al (2019) A comparative assessment of flood susceptibility modeling using multi-criteria decision-making analysis and machine learning methods. J Hydrol 573:311–323

Kool JT, Moilanen A, Treml EA (2013) Population connectivity: recent advances and new perspectives. Landscape Ecol 28(2):165–185

Kumari I, Pradhan L, Chatterjee S (2024) Challenges and strategies in elephant conservation: a comprehensive review of land use impact and management approaches in Dalma Wildlife Sanctuary, India. J Landsc Ecol 17:80–96

Laurance WF, Camargo JL, Luizão RC, Laurance SG, Pimm SL, Bruna EM et al (2011) The fate of Amazonian forest fragments: a 32-year investigation. Biol Cons 144(1):56–67
Laurance WF, Sayer J, Cassman KG (2014) Agricultural expansion and its impacts on tropical nature. Trends Ecol Evol 29(2):107–116
Lindenmayer DB, Fischer J (2013) Habitat fragmentation and landscape change: an ecological and conservation synthesis. Island Press
Lindenmayer DB, Fischer J, Felton A, Crane M, Michael D, Macgregor C et al (2008) Novel ecosystems resulting from landscape transformation create dilemmas for modern conservation practice. Conserv Lett 1(3):129–135
Magnago LFS, Edwards DP, Edwards FA, Magrach A, Martins SV, Laurance WF (2014) Functional attributes change but functional richness is unchanged after fragmentation of Brazilian Atlantic forests. J Ecol 102(2):475–485
Magrach A, Laurance WF, Larrinaga AR, Santamaria L (2014) Meta-analysis of the effects of forest fragmentation on interspecific interactions. Conserv Biol 28(5):1342–1348
Mallick J (2016) Geospatial-based soil variability and hydrological zones of Abha semi-arid mountainous watershed, Saudi Arabia. Arab J Geosci 9(4):281
Mallick PH, Chakraborty SK (2018) Forest, wetland and biodiversity: revealing multi-faceted ecological services from ecorestoration of a degraded tropical landscape. Ecohydrol Hydrobiol 18(3):278–296
Mandal M, Chatterjee ND (2021) Forest landscape and its ecological quality: a stepwise spatiotemporal evaluation through patch-matrix model in Jhargram District, West Bengal State, India. Reg Sustain 2(2):164–176
Mandal M, Chatterjee ND (2023) Geo-spatial analysis of forest landscape for wildlife management. Springer
Matte ALL, Müller SC, Becker FG (2015) Forest expansion or fragmentation? Discriminating forest fragments from natural forest patches through patch structure and spatial context metrics. Austral Ecol 40(1):21–31
McGarigal K (1995) FRAGSTATS: spatial pattern analysis program for quantifying landscape structure, vol 351. US Department of Agriculture, Forest Service, Pacific Northwest Research Station
McRae BH, Dickson BG, Keitt TH, Shah VB (2008) Using circuit theory to model connectivity in ecology, evolution, and conservation. Ecology 89(10):2712–2724
Metzger JP, Martensen AC, Dixo M, Bernacci LC, Ribeiro MC, Teixeira AMG, Pardini R (2009) Time-lag in biological responses to landscape changes in a highly dynamic Atlantic Forest region. Biol Cons 142(6):1166–1177
Meyfroidt P, Carlson KM, Fagan ME, Gutiérrez-Vélez VH, Macedo MN, Curran LM et al (2014) Multiple pathways of commodity crop expansion in tropical forest landscapes. Environ Res Lett 9(7):074012
Mitra S, Mishra P, Daniel JA, Balaji S (2019) Smart light for home with automatic direction and intensity adjustment using arduino. Int J Recent Technol Eng (IJRTE) 8(1S4):479–484
Mukhopadhyay A, Mitra D, Hazra S (eds) (2021) Sundarbans Mangrove systems: a geo-informatics approach. CRC Press
Newmark WD, Jenkins CN, Pimm SL, McNeally PB, Halley JM (2017) Targeted habitat restoration can reduce extinction rates in fragmented forests. Proc Natl Acad Sci 114(36):9635–9640
Penjor U, Jamtsho R, Sherub S (2022) Anthropogenic land-use change shapes bird diversity along the eastern Himalayan altitudinal gradient. J Appl Ecol 59(3):847–859
Pettorelli N (2013) The normalized difference vegetation index. Oxford University Press
Pfeifer M, Lefebvre V, Peres CA, Banks-Leite C, Wearn OR, Marsh CJ et al (2017) Creation of forest edges has a global impact on forest vertebrates. Nature 551(7679):187–191
Qiu J, Carpenter SR, Booth EG, Motew M, Kucharik CJ (2020) Spatial and temporal variability of future ecosystem services in an agricultural landscape. Landscape Ecol 35(11):2569–2586

Ramesh T, Sankar K, Qureshi Q, Kalle R (2012) Group size and population structure of megaherbivores (gaur Bos gaurus and Asian elephant Elephas maximus) in a deciduous habitat of Western Ghats, India. Mammal Study 37(1):47–54

Randhawa K, Bhattacharya P, Shimrah T (2024) Spatio-temporal analysis of urban forest fragmentation in Delhi, India, using landscape fragmentation tool v2. 0. Environ Monit Assess 196(12):1273

Rempel RS, Kaukinen D, Carr AP (2012) Patch analyst and patch grid. Ontario Ministry of Natural Resources, Centre for Northern Forest Ecosystem Research, Thunder Bay, Ontario

Roy M, Sukumar R (2017) Railways and wildlife: a case study of train-elephant collisions in northern West Bengal, India. Railway ecology. Springer International Publishing, Cham, pp 157–177

Rudel TK, Defries R, Asner GP, Laurance WF (2009) Changing drivers of deforestation and new opportunities for conservation. Conserv Biol 23(6):1396–1405

Sahoo KP, Roy A, Mandal MH, Yasmin B, Bhattacharjee S, Siddique G (2023) Appraisal of coexistence and interdependence of forest and tribes in Jhargram District of West Bengal, India using SWOT–AHP analysis. GeoJournal 88(2):1493–1513

Schlaepfer DR, Braschler B, Rusterholz HP, Baur B (2018) Genetic effects of anthropogenic habitat fragmentation on remnant animal and plant populations: a meta-analysis. Ecosphere 9(10):e02488

Shaffer LJ, Khadka KK, Van Den Hoek J, Naithani KJ (2019) Human-elephant conflict: a review of current management strategies and future directions. Front Ecol Evol 6:235

Sharma M, Areendran G, Raj K, Sharma A, Joshi PK (2016) Multitemporal analysis of forest fragmentation in Hindu Kush Himalaya—a case study from Khangchendzonga Biosphere Reserve, Sikkim, India. Environ Monit Assess 188(10):596

Stupariu MS, Cushman SA, Pleşoianu AI, Pătru-Stupariu I, Fuerst C (2022) Machine learning in landscape ecological analysis: a review of recent approaches. Landscape Ecol 37(5):1227–1250

Sudhakar Reddy C, Jha CS, Dadhwal VK, Hari Krishna P, Vazeed Pasha S, Satish KV et al (2016) Quantification and monitoring of deforestation in India over eight decades (1930–2013). Biodivers Conserv 25(1):93–116

Taubert F, Fischer R, Groeneveld J, Lehmann S, Müller MS, Rödig E et al (2018) Global patterns of tropical forest fragmentation. Nature 554(7693):519–522

Theobald DM, Crooks KR, Norman JB (2011) Assessing effects of land use on landscape connectivity: loss and fragmentation of western US forests. Ecol Appl 21(7):2445–2458

Tiwary M (2019) Participatory forest policies and politics in India: joint forest management institutions in Jharkhand and West Bengal. Routledge

Vogt P, Riitters KH, Estreguil C, Kozak J, Wade TG, Wickham JD (2007) Mapping spatial patterns with morphological image processing. Landscape Ecol 22(2):171–177

Yang L, Driscol J, Sarigai S, Wu Q, Chen H, Lippitt CD (2022) Google Earth Engine and artificial intelligence (AI): a comprehensive review. Remote Sens 14(14):3253

Zeller KA, McGarigal K, Whiteley AR (2012) Estimating landscape resistance to movement: a review. Landscape Ecol 27(6):777–797

Chapter 4
Forest Biomass and Carbon Stock

Abstract This study presents a comprehensive assessment of forest biomass and carbon stock distribution across the Jangal Mahal region of West Bengal, India, employing integrated remote sensing and field-based methodologies to quantify ecosystem carbon storage potential. Utilizing Sentinel-2 multispectral imagery combined with Random Forest machine learning algorithms, the research analyzed 130 systematically distributed field plots (30 × 30 m) across 2709 km^2 of tropical deciduous forest landscape. Field measurements encompassed tree enumeration, diameter at breast height, and height data, processed through Chave et al.'s allometric equations to determine above-ground biomass and carbon stocks. The weighted extraction approach demonstrated superior performance over centroid-based methods, achieving $R^2 = 0.48$ and RMSE = 25.36% in biomass prediction models. Results revealed substantial spatial heterogeneity, with above-ground biomass ranging from 37.4 to 268.6 tons/ha and exhibiting pronounced skewness toward lower productivity classes (47.58% < 114.4 tons/ha). Carbon stock distribution displayed a distinctive bimodal pattern, with 44.34% of areas containing <25 tons C/ha and 33.46% exceeding 95 tons C/ha, indicating polarized landscape conditions. Comparative analysis across forest types revealed the ecological superiority of native Sal forests (154.36 ± 17.82 tons/ha) over exotic plantations, with Acacia achieving only 67.5% and Eucalyptus 57% of indigenous forest productivity. Mixed forests demonstrated intermediate performance (129.41 tons/ha), suggesting species diversification strategies can partially offset monocultural limitations. These findings provide critical baseline data for REDD + implementation, climate mitigation planning, and evidence-based forest management in tropical deciduous ecosystems, emphasizing the irreplaceable value of indigenous forest conservation for regional carbon sequestration and biodiversity preservation initiatives.

Keywords Forest biomass estimation · Carbon stock assessment · Remote sensing and GIS · Random forest model · Jangal Mahal

P. K. Shit and M. Hasanuzzaman, *Forest Landscape Dynamics in West Bengal*, Advances in Asian Human-Environmental Research,
https://doi.org/10.1007/978-3-032-23700-2_4

4.1 Introduction

Forest biomass represents the total mass of living organic matter within forest ecosystems, encompassing above-ground components including tree stems, branches, foliage, and below-ground root systems, as well as associated understory vegetation and organic soil layers (Nair et al. 2022). This biological reservoir constitutes one of the largest terrestrial carbon pools on Earth, storing approximately 861 gigatons of carbon globally, which represents roughly 45% of terrestrial carbon stocks and plays a fundamental role in regulating atmospheric CO_2 concentrations and global climate systems (Pan et al. 2011; Friedlingstein et al. 2020). The quantification of forest biomass has gained unprecedented importance in contemporary environmental science due to its critical role in climate change mitigation strategies, carbon trading mechanisms, and international policy frameworks such as the Reducing Emissions from Deforestation and forest Degradation (REDD+) initiative under the United Nations Framework Convention on Climate Change (Köhl et al. 2009).

The relationship between forest biomass and carbon content follows well-established biogeochemical principles, with carbon typically comprising 45–50% of dry biomass weight in most tree species, though this proportion can vary significantly across taxonomic groups, forest types, and environmental conditions (Thomas and Martin, 2012; Martin et al. 2018). Above-ground biomass (AGB) generally accounts for 70–80% of total forest carbon in temperate and tropical forests, while below-ground biomass, primarily root systems, contributes the remaining 20–30%, with these proportions varying considerably based on forest age, species composition, soil conditions, and climatic factors (Cairns et al. 1997; Mokany et al. 2006). The temporal dynamics of carbon accumulation in forest ecosystems follow predictable patterns related to stand development, with young forests typically exhibiting rapid carbon sequestration rates during early growth phases, while mature forests maintain relatively stable carbon stocks with slower net accumulation rates, though recent research suggests that old-growth forests may continue to sequester carbon at significant rates over extended periods (Luyssaert et al. 2008; Stephenson et al. 2014).

Spatial heterogeneity in forest carbon stocks reflects complex interactions between climatic variables, soil properties, topographic characteristics, species composition, and disturbance regimes, creating intricate patterns across multiple scales from local forest stands to continental biomes. Tropical forests, despite occupying only 7% of global land surface, contain approximately 25% of terrestrial carbon stocks, with the Amazon basin alone storing an estimated 150–200 gigatons of carbon, highlighting the disproportionate importance of these ecosystems in global carbon cycling (Saatchi et al. 2011; Baccini et al. 2017). Boreal forests, characterized by slower growth rates but extensive coverage across northern latitudes, contribute significantly to global carbon storage through both biomass accumulation and substantial soil organic carbon pools, with permafrost regions containing additional carbon reserves vulnerable to climate-induced thawing (Bradshaw and Warkentin 2015; Schuur et al. 2015). Temperate forests, while generally exhibiting intermediate carbon densities, demonstrate high variability related to management practices, with managed forests typically storing 20–30% less carbon than natural forests, though sustainable management practices can optimize carbon sequestration while maintaining timber production (Nunery and Keeton 2010; Griscom et al. 2017).

Forest carbon dynamics are fundamentally influenced by disturbance regimes, including natural processes such as wildfires, insect outbreaks, windstorms, and drought events, as well as anthropogenic activities including deforestation, selective logging, and land-use conversion. These disturbances can rapidly transform forests from carbon sinks to carbon sources, with immediate emissions from biomass combustion or decomposition, followed by complex recovery trajectories that depend on regeneration patterns, species succession, and subsequent management interventions (Kurz et al. 2008). Climate change amplifies these disturbance effects through increased fire frequency and intensity, extended drought periods, and shifting pest and pathogen distributions, creating feedback loops that may fundamentally alter forest carbon storage capacity and distribution patterns across landscapes (Allen et al. 2010; Seidl et al. 2017).

The methodological approaches for quantifying forest carbon stocks have evolved from traditional forest inventory techniques based on allometric equations relating tree diameter and height to biomass, toward sophisticated integration of remote sensing technologies, process-based modeling, and machine learning algorithms. Contemporary carbon assessment protocols incorporate multiple scales of observation, from individual tree measurements through plot-level sampling to landscape-scale remote sensing applications, with each approach contributing unique advantages and limitations to comprehensive carbon monitoring systems (Chave et al. 2014; Duncanson et al. 2019). Uncertainty quantification has become increasingly important in carbon stock assessments, particularly for applications in carbon markets and national greenhouse gas inventories, requiring careful consideration of measurement errors, sampling variability, allometric model uncertainty, and temporal dynamics in carbon pools (Réjou-Méchain et al. 2014; Marvin et al. 2014).

4.2 Forest Biomass and Carbon Stock in West Bengal

West Bengal, positioned in the eastern Indian subcontinent between 21°25' N and 27°13' N latitude and 85°50' E to 89°50' E longitude, encompasses approximately 88,752 km^2 and supports diverse forest ecosystems that play a crucial role in regional carbon cycling and climate regulation (Das 2021). The state's forest cover extends across 16,902 km^2, representing 19.04% of the total geographical area, with these forests distributed across distinct ecological zones including the tropical moist deciduous forests of the Gangetic Plains, subtropical hill forests of the Darjeeling Himalayas, and mangrove ecosystems of the Sundarbans Delta (Forest Survey of India 2021). The heterogeneous topography and climatic gradients, ranging from sea level in the coastal regions to over 3600 m in the northern hills, create diverse habitat conditions that support varying forest types with distinctly different biomass accumulation patterns and carbon storage capacities (Nandy et al. 2011).

The Sundarbans mangrove forests, covering approximately 4200 km^2 within West Bengal and representing the world's largest contiguous mangrove ecosystem, constitute a unique and highly productive carbon reservoir with distinctive biogeochemical characteristics (Mukhopadhyay et al. 2021). These tidal forests demonstrate exceptionally high-carbon storage potential, with above-ground biomass estimates ranging from 150 to 300 Mg ha^{-1} and total ecosystem carbon stocks (including below-ground and soil carbon) reaching 400-600 Mg C ha^{-1}, significantly higher than many terrestrial forest systems due to the anaerobic soil conditions that reduce decomposition rates and promote long-term carbon sequestration (Ghosh et al. 2015; Ray et al. 2011). The dominant species including Heritiera fomes, Excoecaria agallocha, Ceriops decandra, and Avicennia officinalis exhibit specialized adaptations for saline environments and demonstrate rapid biomass accumulation rates, with annual carbon sequestration rates estimated at 2.5-4.8 Mg C ha^{-1} $year^{-1}$, making these ecosystems critical for regional climate change mitigation strategies (Kothamasi et al. 2006; Mandal and Naskar 2008).

The tropical moist deciduous forests of the plains and foothills, predominantly found in districts including Purulia, Bankura, West Midnapore, and Birbhum, are characterized by mixed species composition dominated by Shorea robusta (sal), Terminalia species, Lagerstroemia parviflora, and Madhuca longifolia, with above-ground biomass ranging from 80 to 180 Mg ha^{-1} and carbon stocks varying from 40 to 90 Mg C ha^{-1} depending on forest density, age structure, and management practices (Paul and Das 2025; Murthy et al. 2013). These forests experience significant anthropogenic pressure through selective logging, fuelwood collection, and grazing activities, which substantially impact carbon accumulation patterns and create heterogeneous landscape mosaics with varying degrees of forest degradation and recovery. Research conducted across multiple forest divisions indicates that well-preserved sal forests can achieve carbon stocks comparable to tropical evergreen forests, while degraded areas exhibit 40–60% lower carbon storage capacity, highlighting the importance of conservation and restoration initiatives for maintaining ecosystem carbon services (Chitale et al. 2014; Sen et al. 2012).

The subtropical and temperate forests of the Darjeeling hills, extending across elevational gradients from 200 to 3600 m, encompass diverse forest types including subtropical broadleaved forests dominated by Castanopsis and Quercus species at lower elevations, mixed coniferous forests with Cryptomeria japonica and Pinus species at mid-elevations, and alpine scrub vegetation at higher altitudes (Tolangay and Moktan 2024). These montane forests exhibit carbon storage patterns strongly influenced by elevation, aspect, and slope characteristics, with above-ground biomass ranging from 120 to 250 Mg ha^{-1} in dense broadleaved forests to 80-150 Mg ha^{-1} in coniferous stands (Sharma et al. 2010; Baishya and Barik 2011). The cooler temperatures and higher precipitation at elevated sites promote slower decomposition rates and enhanced soil organic carbon accumulation, with total ecosystem carbon stocks often exceeding 200 Mg C ha^{-1}, particularly in undisturbed old-growth forest patches that serve as important refugia for biodiversity conservation and carbon storage (Usuga et al. 2010).

Anthropogenic impacts on forest biomass and carbon stocks in West Bengal reflect complex interactions between population pressure, agricultural expansion, infrastructure development, and forest management policies, with deforestation rates estimated at 0.2-0.5% annually across different forest types and regions (Reddy et al. 2013; Joshi et al. 2006). The implementation of Joint Forest Management (JFM) programs since the 1990s has demonstrated positive impacts on forest recovery and carbon sequestration, with community-managed forests showing 15-25% higher biomass accumulation rates compared to conventionally managed areas, though success varies significantly based on local socio-economic conditions and community participation levels (Saha and Sundriyal 2012; Angelsen 2009). Climate change projections for the region suggest potential shifts in forest composition and productivity, with increasing temperatures and altered precipitation patterns likely to affect carbon storage capacity, particularly in the vulnerable Sundarbans ecosystem, where sea-level rise poses additional threats to mangrove carbon stocks (Chanda and Akhand 2023).

Recent advances in remote sensing applications and GIS-based modeling have enhanced the precision of forest carbon assessments in West Bengal, with studies integrating Landsat, MODIS, and Sentinel satellite data with field measurements to develop spatially explicit carbon maps across the state. These technological approaches have revealed significant spatial heterogeneity in carbon distribution, with hotspots of high-carbon density concentrated in the Sundarbans core areas, protected forest zones in the Darjeeling hills, and well-preserved sal forest patches in the western districts (Nandy et al. 2011; Qasim et al. 2013). Machine learning algorithms, particularly Random Forest and Support Vector Machine techniques, have shown promise for scaling up field-based biomass estimates to landscape levels, achieving accuracy levels of 80–85% when validated against independent ground-truth data, though challenges remain in accurately capturing the complex structural characteristics of mangrove forests and mixed species compositions in deciduous forest systems (Thapa et al. 2023).

4.3 Forest Biomass and Carbon Stock Calculation Tools and Techniques in GIS

The integration of Geographic Information Systems (GIS) with remote sensing technologies has revolutionized forest biomass and carbon stock estimation, providing researchers and forest managers with sophisticated tools for large-scale environmental monitoring and carbon accounting. Contemporary GIS-based approaches leverage multiple data sources, including satellite imagery, LiDAR data, and field measurements, to develop robust models for quantifying forest carbon dynamics across diverse spatial and temporal scales. The foundation of these methodologies' rests on allometric equations that establish relationships between easily measurable tree parameters and biomass content, which are then scaled up using geospatial analysis techniques (Chave et al. 2014; Duncanson et al. 2019).

Remote sensing integration within GIS platforms enables the development of predictive models through various spectral indices and biophysical parameters derived from multispectral and hyperspectral imagery. The Normalized Difference Vegetation Index (NDVI), Enhanced Vegetation Index (EVI), and Leaf Area Index (LAI) serve as fundamental inputs for biomass estimation algorithms, while more advanced techniques incorporate radar backscatter coefficients from Synthetic Aperture Radar (SAR) data to penetrate forest canopies and assess structural characteristics (Santoro et al. 2021; Rodríguez-Veiga et al. 2016). Machine learning algorithms, particularly Random Forest, Support Vector Machines, and neural networks, have emerged as powerful tools for processing these multi-dimensional datasets within GIS environments, enabling the development of spatially explicit biomass maps with improved accuracy compared to traditional statistical approaches (Baccini et al. 2017; Xu et al. 2021).

LiDAR technology integration represents a paradigm shift in forest biomass estimation, providing three-dimensional structural information that directly correlates with forest carbon stocks. Airborne Laser Scanning (ALS) and terrestrial laser scanning generate point clouds that, when processed through GIS-based algorithms, yield precise measurements of canopy height, crown dimensions, and vertical forest structure (Asner et al. 2013). The Global Ecosystem Dynamics Investigation (GEDI) spaceborne LiDAR mission has further expanded these capabilities to global scales, providing unprecedented opportunities for continental and global carbon stock assessments through GIS-based processing workflows (Dubayah et al. 2020). These LiDAR-derived metrics are integrated with field-based allometric equations to create spatially continuous biomass maps, with uncertainty quantification becoming increasingly important for policy applications and carbon market mechanisms.

The development of standardized protocols and tools within GIS platforms has facilitated widespread adoption of these techniques across research institutions and environmental agencies. Software packages such as the Forest Carbon Assessment and Monitoring (FCAM) toolkit, the Integrated Land and Water Information System (ILWIS), and specialized R packages like "BIOMASS" and "rLiDAR" provide researchers with accessible frameworks for implementing complex biomass estimation workflows (Réjou-Méchain et al. 2017; Silva et al. 2017). Tools such as the InVEST Carbon Model and the FAO's Collect Earth integrate GIS to automate carbon stock assessments, incorporating land-use change dynamics and temporal trends. Cloud-based platforms, particularly Google Earth Engine, have democratized access to large-scale geospatial analysis capabilities, enabling researchers to process vast archives of satellite imagery for temporal biomass change detection and forest carbon monitoring without requiring extensive computational infrastructure (Gorelick et al. 2017; Hansen et al. 2013).

Quality assessment and validation remain critical components of GIS-based biomass estimation, requiring careful consideration of uncertainty propagation from field measurements through remote sensing calibration to final carbon stock maps. Cross-validation techniques, independent ground-truth datasets, and inter-comparison studies help establish confidence intervals and identify sources of systematic bias in GIS-derived estimates (Réjou-Méchain- et al. 2014). The integration of multiple sensors and methodologies through ensemble modeling approaches has shown promise for reducing uncertainty and improving the reliability of carbon stock assessments, particularly important for applications in REDD+ mechanisms and national greenhouse gas inventories (Avitabile et al. 2016; Santoro et al. 2015).

4.4 A Case Study on Forest Biomass and Carbon Stock in the Jangal Mahal of West Bengal

The Jangal Mahal region of West Bengal represents a critical ecological zone where tropical deciduous forests serve as significant carbon reservoirs within the Indian subcontinent's climate mitigation framework. This forested landscape, encompassing districts such as Purulia, Bankura, West Midnapore, Jhargram, and Birbhum, constitutes approximately 35% of West Bengal's total forest area and demonstrates remarkable potential for carbon sequestration and biomass accumulation (Das 2021). The region's forest ecosystems, predominantly characterized by tropical dry deciduous vegetation, harbor substantial above-ground biomass ranging from 50 to 250 tons per hectare, with corresponding carbon stocks varying between 25 and 125 tons C ha^{-1}, reflecting the heterogeneous nature of forest stand compositions and successional stages (Paul and Das 2025).

Recent comprehensive assessments utilizing non-destructive allometric equations have revealed that the Jangal Mahal forests maintain an average biomass density of 180–220 tons ha^{-1} in mature stands, significantly contributing to regional carbon storage capacity (Karmakar et al. 2023). The systematic evaluation of Indian forest ecosystems indicates that tropical deciduous forests, similar to those in Jangal Mahal, demonstrate substantial carbon sequestration potential, with annual carbon accumulation rates ranging from 2.5 to 8.0 tons C ha^{-1} $year^{-1}$ depending on forest age, species composition, and anthropogenic disturbance levels (Shit et al. 2020). The application of species-specific allometric models has enhanced biomass estimation accuracy by 15-25% compared to generalized equations, particularly for dominant species such as Shorea robusta, Terminalia arjuna, and Anogeissus latifolia that characterize the region's forest canopy structure (Dutta Roy et al. 2023).

Geospatial analysis combined with field-based biomass data indicates substantial ecosystem dynamics, with biomass analysis revealing considerable variation across forest patches, emphasizing the need for landscape-level conservation strategies. The carbon storage potential of these forests extends beyond above-ground components, with soil organic carbon contributing an additional 40–60 tons C ha^{-1} in the upper 30 cm soil profile, establishing the total ecosystem carbon stock at approximately 100–185 tons C ha^{-1} (Lorenz and Lal 2018). Studies on tropical forests in West Bengal demonstrate that these ecosystems can sequester significant amounts of atmospheric carbon, with the estimation of above-ground biomass being crucial for quantifying carbon stock and understanding the effects of deforestation on global carbon balance.

The implementation of REDD+ mechanisms in the Jangal Mahal region requires precise biomass quantification protocols that integrate remote sensing technologies with ground-truth validation to ensure accurate carbon credit calculations (Rajput et al. 2025). Advanced allometric equation development shows that species-specific models combining diameter with height measurements provide more precise above-ground biomass predictions, essential for developing robust monitoring, reporting, and verification systems. Forest degradation analysis indicates that anthropogenic pressures have reduced biomass density by 20-35% in disturbed areas compared to protected forest patches, highlighting the urgent need for sustainable forest management practices that balance community needs with conservation objectives (Mukherjee et al. 2017). The integration of community-based forest management approaches with scientific carbon stock assessments provides a comprehensive framework for enhancing both biodiversity conservation and climate change mitigation in this ecologically vital region of eastern India.

The research focuses on a strategically selected study area within the Jangal Mahal region of West Bengal, India, specifically encompassing the forest-dominated districts of Jhargram and Paschim Medinipur (Shit et al. 2020). This geographical selection directly aligns with the primary research objectives, as these districts represent critical forest ecosystems experiencing significant environmental pressures. The entire study area (Fig. 4.1) serves as the focal point for this chapter, providing a comprehensive framework for examining forest compositional changes and degradation processes that pose substantial threats to local ecosystem integrity and carbon sequestration potential (Shit et al. 2020).

The selection of this region for intensive investigation was strategically determined based on its ecological significance and the urgent need to evaluate carbon pool dynamics alongside other critical environmental parameters (Roy et al. 2020). This focused approach enables a thorough assessment of the multifaceted challenges confronting both the diverse ecological communities and the human populations dependent on these forest resources. The study area's designation facilitates comprehensive analysis of ecosystem services, biodiversity conservation needs, and sustainable forest management requirements within this ecologically vital region of eastern India (Shit et al. 2020).

4.4.1 *Material and Methods*

Field Data Collection Methodology

Field measurements were conducted across 130 systematically distributed sampling sites, with each site comprising a standardized 30 × 30-m square plot. Within these established plots, comprehensive biometric data were recorded for all individual trees present.

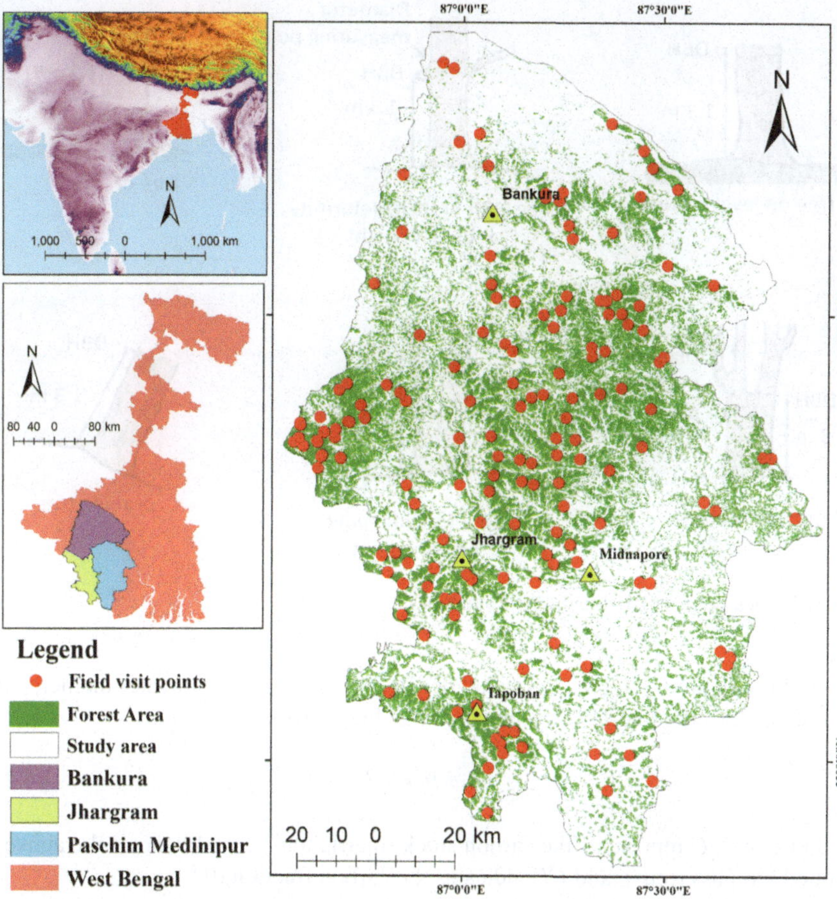

Fig. 4.1 Location map of the study area

The collected dataset encompassed tree enumeration, diameter at breast height (DBH), and individual tree heights (Fig. 4.2). Geographic positioning data were simultaneously captured for each plot, including precise coordinates for both internal reference points and corner boundaries, utilizing a portable GPS unit for spatial accuracy. Standardized DBH measurements were taken at 1.3 m above-ground level for each tree specimen within the sampling plots. Above-ground biomass (AGB) calculations were performed using the enhanced allometric equation developed by Chave et al. (2014) (Eq. 4.1). Carbon stock estimates were subsequently derived through the carbon fraction methodology established by Hirata et al. (2014) (Eq. 4.2), with the analytical framework presented in Fig. 4.3.

$$\mathrm{AGB} = 0.0673\ (\rho \times D2 \times H)\ 0.976 \tag{4.1}$$

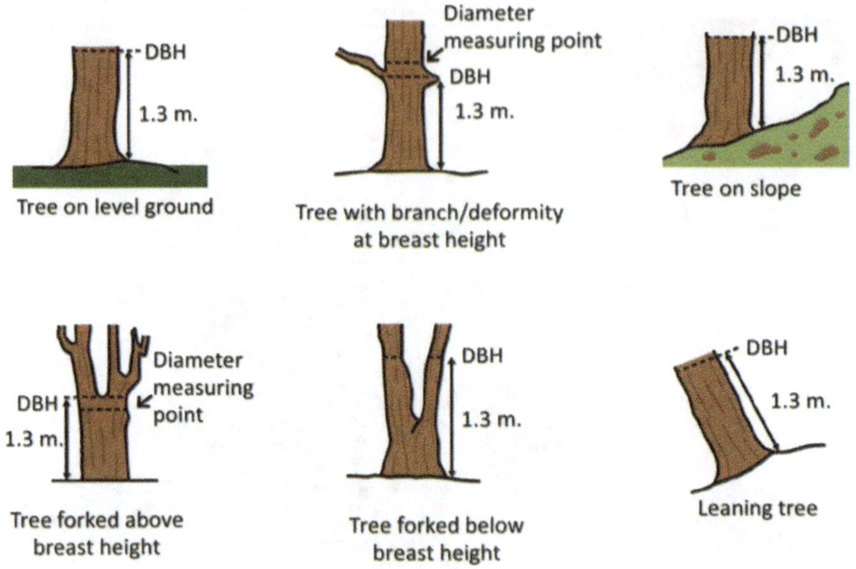

Fig. 4.2 Measurement DBH in different situations

In this equation, ρ represents wood density (g cm^{-3}), DBH indicates diameter at breast height (cm), and H corresponds to tree height (m).

$$C = B \times CF \tag{4.2}$$

In Eq. 4.2, C represents the carbon stock (metric tons), AGB denotes the aboveground biomass (tons), and CF indicates the carbon fraction (0.5).

The quantification of AGB and carbon stock relies on three fundamental biometric parameters: wood density, tree diameter, and height. These equations provide a robust analytical framework for transforming field measurements into quantitative assessments of tree biomass and carbon storage within the surveyed plots.

Sentinel-2 Data Processing

This study employed Sentinel-2A and Sentinel-2B satellite imagery acquired between September 15, 2023, and December 25, 2024, sourced from the USGS Earth Explorer platform (https://earthexplorer.usgs.gov). Image selection was based on optimal cloud cover conditions and temporal alignment with field sampling campaigns. The Level-1C (L1C) products deliver top-of-atmosphere (TOA) reflectance data with ortho-rectified geometry and sub-pixel accuracy through systematic geometric correction. The multispectral dataset encompasses 13 spectral bands spanning visible, near-infrared, red-edge, and shortwave infrared wavelengths.

Fig. 4.3 **a** Measurement of tree girth using a diameter tape at breast height (DBH), demonstrating standard forest inventory practice in a Sal-dominated stand. **b** Field investigator recording trunk circumference on a mature tree to estimate biomass and structural characteristics. **c** Close-range girth measurement of a standing tree within the forest plot, illustrating systematic data collection for vegetation analysis. **d** Another example of DBH measurement in a dry deciduous forest, showing field methodology used for assessing tree growth and forest condition

For this analysis, 10 bands from the Sentinel-2A MSI were selected, excluding bands 1, 9, and 10 due to their specific focus on water vapor and atmospheric correction applications rather than vegetation analysis (Table 4.1).

Following data acquisition, the ortho-rectified Sentinel-2A and 2B L1C imagery underwent atmospheric correction processing to generate bottom-of-atmosphere

Table 4.1 Summary of input parameters employed in above-ground biomass modeling, comprising raw spectral bands, texture characteristics, and established vegetation indices derived from Sentinel-2A multispectral imagery

Variables	Equation	References
Original spectral bands		European space agency, Sentinel-2 user handbook, ESA standard document: European space agency; Paris, France (2015)
Blue	B2 10 m	
Red	B4 10 m	
Red edge 1	B5 20 m	
Red edge 2	B6 20 m	
Red edge 3	B7 20 m	
NIR	B8 10 m	
NIR narrow	B8A 20 m	
SWIR	B11 20 m	
SWIR	B12 20 m	
Traditional spectral indices		
DVI	$\rho\text{NIR} - \rho\text{Red}$	Tucker (1979)
EVI	$2.5(p\text{NIR} - \rho\text{Red})/(p\text{NIR}) -6\rho\text{Red} - 7.5\rho\text{Blue} + 1)$	Matsushita et al. (2007)
NDVI	$(p\text{NIR} - \rho\text{Red})/(p\text{NIR} + \rho\text{Red})$	Jw (1973)
Gray-level co-occurrence matrix (GLCM)		
Variance (VAR)	$\sum_{i,j=0}^{N-1} pi, j/(1 - ui)$	Haralick et al. (2007)
Homogeneity (HOM)	$\sum_{i,j=0}^{N-1} ipi, j/(1 + (1 - j)^2)$	
Contrast (CON)	$\sum_{i,j=0}^{N-1} iPi, j(i - j)$	
Dissimilarity (DIS)	$\sum_{i,j=0}^{N-1} iPi, j(i - j)$	
Correlation (COR)	$\sum_{i,j=0}^{N-1} i \sum_{i,j=0}^{N-1} (ijpi, j - uiui) /\sigma_i^2\sigma_i^2$	

(BOA) reflectance products using the Sen2Cor version 2.8 plugin from the European Space Agency (ESA). The preprocessing workflow included image mosaicking to create seamless, continuous coverage across the study area, followed by linear smoothing algorithms to reduce noise and enhance analytical quality. Spatial resampling was performed using ESA's SNAP version 7.0 to standardize all Sentinel-2A and 2B bands to 10-m resolution from their original 20-m pixel size. To assess the potential of Sentinel-2 MSI data for above-ground biomass estimation in forest ecosystems, multiple vegetation indices were calculated from the processed imagery mosaics. These spectral indices, including the Enhanced Vegetation Index (EVI), exploit specific wavelength combinations to quantify vegetation density and health across the landscape (Naidoo et al. 2019; Sibanda et al. 2015). Textural feature extraction was performed on the EVI imagery to derive four key spatial metrics: entropy, contrast, correlation, and dissimilarity (Haralick et al. 2007; Kelsey and Neff 2014; Dong et al. 2020). The texture analysis employed Gray Level Co-occurrence Matrix (GLCM) methodology to quantify spatial heterogeneity patterns within the imagery. These four texture parameters were selected based on their demonstrated efficacy in above-ground biomass estimation applications.

Window size optimization was critical for texture computation, with sizes determined through correlation analysis between texture-derived estimates and field-observed AGB values, informed by established literature benchmarks. Texture metrics were calculated using four distinct processing window dimensions (3×3, 5×5, 7×7, and 9×9 pixels) with 64-level gray-scale quantization (Kelsey and Neff 2014; Das 2021). This integrated analytical framework leverages the distinctive spectral characteristics of Sentinel-2A and 2B MSI sensors, combining multispectral imagery, vegetation indices, and spatial texture analysis to enhance AGB estimation accuracy in forest ecosystems. The approach provides valuable insights into forest structural dynamics and ecosystem carbon storage potential.

Machine Learning Algorithm for AGB Modeling: Random Forest Approach

The Random Forest (RF) algorithm represents a robust ensemble learning methodology that enhances predictive performance through the integration of multiple decision trees (Breiman 2001). The RF modeling framework employed bootstrap aggregation (bagging), where each constituent decision tree was trained using random sampling with replacement from the original dataset. Approximately two-thirds of the data were allocated for unpruned tree construction, while the remaining one-third constituted out-of-bag (OOB) samples for internal model validation and error estimation (Prasad et al. 2006).

Variable importance assessment in RF utilizes two primary metrics: percent increase in mean squared error (%IncMSE) and increase in node purity. This study employed the %IncMSE criterion, which quantifies accuracy reduction resulting from random permutation of OOB observations-a widely adopted approach in biomass modeling studies (Li et al. 2019). Higher %IncMSE values indicate greater variable significance in model predictions. RF model optimization requires tuning two critical hyperparameters: (i) ntree, representing the total number of trees in the ensemble, and (ii) mtry, defining the number of variables considered for optimal splitting at each node. Hyperparameter optimization was achieved through grid-search methodology coupled with fivefold cross-validation to minimize root mean square error (RMSE) and enhance model reliability.

Model Validation and Performance Assessment

The complete dataset of 130 biomass samples was randomly partitioned into training (70%) and testing (30%) subsets to ensure robust model evaluation. Cross-validation techniques were implemented to assess model generalization capacity given the limited sample size. Model performance was evaluated using three statistical metrics: root mean square error (RMSE) (Eq. 4.3), coefficient of determination (R^2) (Eq. 4.4), and coefficient of variation of the root mean square error (CV-RMSE) (Eq. 4.5) (Shit et al. 2020; Hasanuzzaman et al. 2024).

These performance indicators were calculated through fivefold cross-validation to provide reliable estimates of model accuracy and stability. Optimal model performance is characterized by high R^2 values approaching unity, alongside minimized RMSE and CV-RMSE values, indicating strong predictive accuracy and reduced prediction uncertainty (Shit et al. 2020).

$$\text{RMSE} = \sqrt{\sum_{i=1}^{n} \frac{\left(y_i - \widehat{y_i}\right)^2}{n}} \tag{4.3}$$

$$\text{R}^2 = 1 - \frac{\sum_{i=1}^{n} \left(y_i - \widehat{y_i}\right)^2}{\sum_{i=1}^{n} \left(y_i - \widehat{y_i}\right)^2} \tag{4.4}$$

$$\text{CV\_RMSE} = \frac{\text{RMSE}}{\overline{y}} \tag{4.5}$$

where y_i and $\widehat{y_i}$ represent the field-measured and predicted AGB values for the ith sample, respectively; $\overline{y}$ denotes the mean observed AGB across all samples; and n indicates the total sample size for each dataset (Fig. 4.4).

4.4.2 Result and Discussion

Field Above-Ground Biomass

The comprehensive field assessment of above-ground biomass (AGB) distribution across the Jangal Mahal region reveals critical insights into the structural composition and carbon storage capacity of this ecologically significant forest landscape. The empirical data demonstrates a pronounced heterogeneity in biomass distribution patterns, with implications for both forest management strategies and climate change mitigation efforts.

The tree density analysis (Fig. 4.5) indicates a substantial forest cover characterized by varying biomass accumulation rates across different spatial units. The distribution pattern suggests that the majority of surveyed areas exhibit moderate-to-high biomass concentrations, with the 61-80 tons per hectare category showing

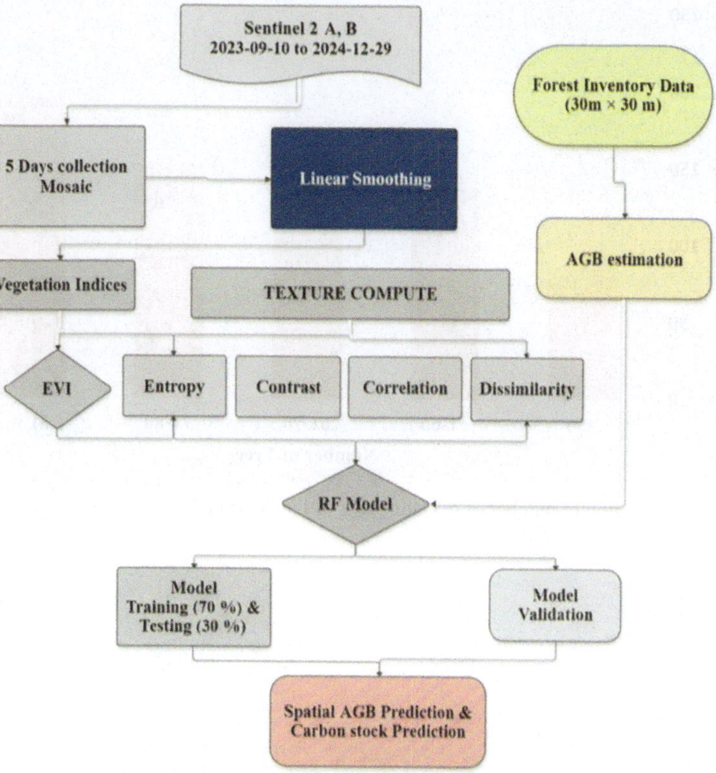

Fig. 4.4 Methodological framework for biomass and carbon stock estimation

the highest frequency of occurrence. This concentration range represents mature forest stands with well-established canopy structures, indicating successful natural regeneration processes and minimal anthropogenic disturbances in these zones.

Particularly noteworthy is the spatial variability evident in Fig. 4.6, where tree density correlates positively with AGB values, demonstrating the fundamental ecological principle that forest productivity increases with stand density up to an optimal threshold. The regions exhibiting biomass values exceeding 81 tons per hectare represent climax forest communities with complex structural arrangements and multi-layered canopies, contributing significantly to regional carbon sequestration capacity.

The field-level AGB distribution across sample plots (Fig. 4.7) reveals important methodological considerations for large-scale forest inventory assessments. The predominant concentration of plots within the 51-100 tons per hectare range suggests that this ecosystem maintains relatively consistent productivity levels across the landscape, with fewer plots exhibiting extreme values. This distribution pattern indicates

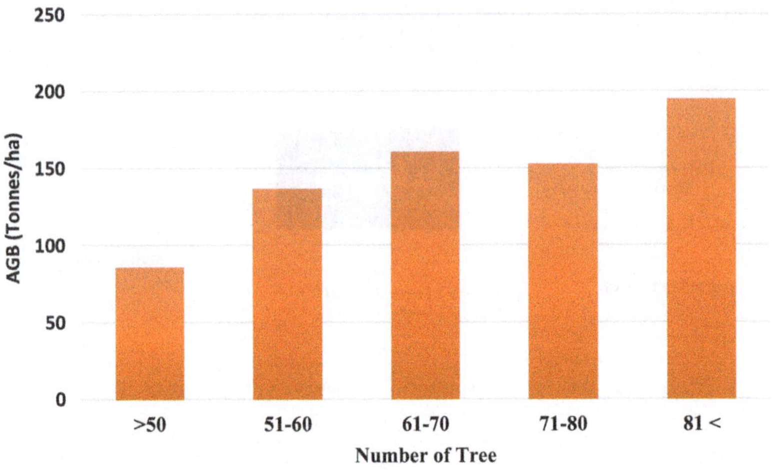

Fig. 4.5 Distribution of number of tree and AGB in Jangal Mahal

Fig. 4.6 Distribution of tree density and AGB in Jangal Mahal

a mature forest system that has achieved relative stability in terms of biomass accumulation, though the presence of plots exceeding 200 tons per hectare suggests microhabitat variations that support exceptionally high productivity.

The data interpretation reveals significant implications for forest conservation strategies in the Jangal Mahal region. The observed biomass distribution indicates that the forest ecosystem maintains substantial carbon storage potential, with the majority of areas contributing meaningfully to climate regulation services. However, the relatively few plots in the highest biomass categories (>200 tons/ha) suggest

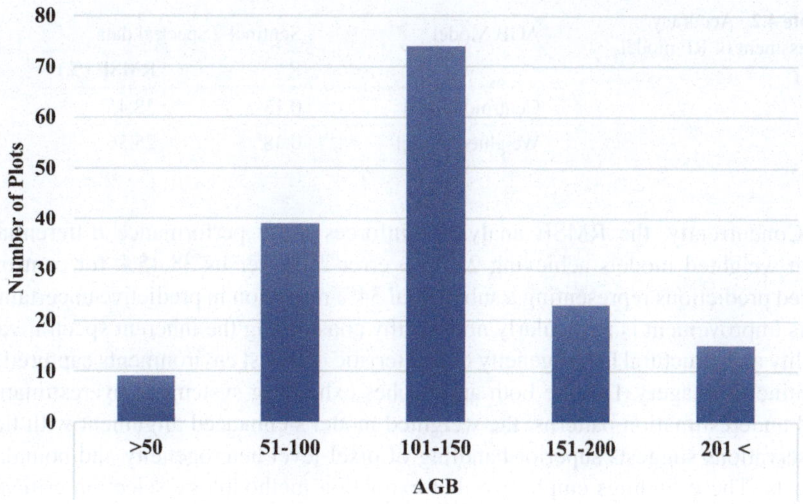

Fig. 4.7 Distribution of field AGB across the sample plots

that these exceptional areas require targeted conservation measures as they represent biodiversity hotspots and maximum carbon storage sites.

From a landscape ecology perspective, the observed patterns suggest successful forest succession processes, with the dominance of medium to high-biomass categories indicating mature forest characteristics. The distribution implies that historical disturbance regimes have been moderate enough to allow forest recovery while maintaining ecological resilience. This finding is particularly significant given the region's history of human settlement and resource utilization.

The research findings contribute valuable baseline data for developing evidence-based forest management protocols and establishing monitoring frameworks for tracking ecosystem health over temporal scales. The biomass distribution patterns provide crucial reference points for assessing future changes in forest structure and function under varying environmental and anthropogenic pressures, ultimately supporting sustainable forest management initiatives in the Jangal Mahal landscape.

Variable Selection and Model Performance Analysis

The comparative analysis of above-ground biomass (AGB) prediction models using Sentinel-2 spectral data reveals profound methodological disparities in remote sensing applications (Table 4.2). The weighted-based variable extraction technique substantially outperforms the centroid-based approach, demonstrating an R^2 improvement from 0.13 to 0.48 nearly quadrupling the model's explanatory capacity. This enhancement signifies that weighted methodology captures approximately 48% of AGB variance compared to merely 13% achieved through centroid extraction, indicating superior integration of spatial complexity within forest ecosystems.

Table 4.2 Accuracy assessment of RF model

AGB Model	Sentinel-2 spectral data	
	R^2	RMSE (%)
Centroid model	0.13	38.45
Weighted model	0.48	25.36

Concurrently, the RMSE analysis reinforces these performance differentials, with weighted models achieving 25.36% error rates versus 38.45% for centroid-based predictions representing a substantial 34% reduction in predictive uncertainty. This improvement is particularly noteworthy considering the inherent spectral variability and structural heterogeneity characteristic of forest environments captured by Sentinel-2 imagery. Despite both approaches exhibiting systematic overestimation and underestimation patterns, the weighted model's enhanced alignment with field observations suggests superior handling of pixel-level heterogeneity and boundary effects. These findings emphasize that extraction methodology selection critically influences model reliability in forest biomass estimation, with direct implications for carbon stock assessments, climate monitoring initiatives, and evidence-based forest management decisions in contemporary environmental research applications.

In contrast, our reported pseudo-R2 of 0.49 and RMSE of 26.62% are generally lower compared to other studies cited. However, compared to conventional field-based methods, machine learning techniques such as Random Forest show significant gains in terms of spatial coverage and reduction in the standard error of AGB estimates from 67 tons/ha to 178 tons/ha. This underscores the capability of these techniques in estimating AGB with considerable accuracy.

The estimated AGB for the Jangal Mahal region of West Bengal aligns with the findings of Roy et al. (2020) and falls within the range of 77-785 Mg/ha reported for tropical montane forests worldwide by González-Jaramillo et al. (2019), which encompassed data from 260 African tropical forests. Despite certain limitations, our study illustrates that using Sentinel-2 imagery combined with RF-based regression methods can effectively facilitate large-scale assessments of forest AGB in tropical rainforest ecosystems.

Predicted Above-Ground Biomass and Carbon Stock

The comprehensive analysis of above-ground biomass (AGB) distribution across the Jangal Mahal region reveals a complex mosaic of forest productivity that reflects the intricate interplay between ecological processes, forest management practices, and environmental gradients. The substantial variation in AGB values, spanning from 37.4 tons/ha to 268.6 tons/ha, indicates a heterogeneous landscape characterized by diverse forest successional stages and varying degrees of anthropogenic influence (Fig. 4.8).

The spatial distribution of above-ground biomass across the 2709 km^2 study area reveals a pronounced skewness toward lower productivity classes, with nearly

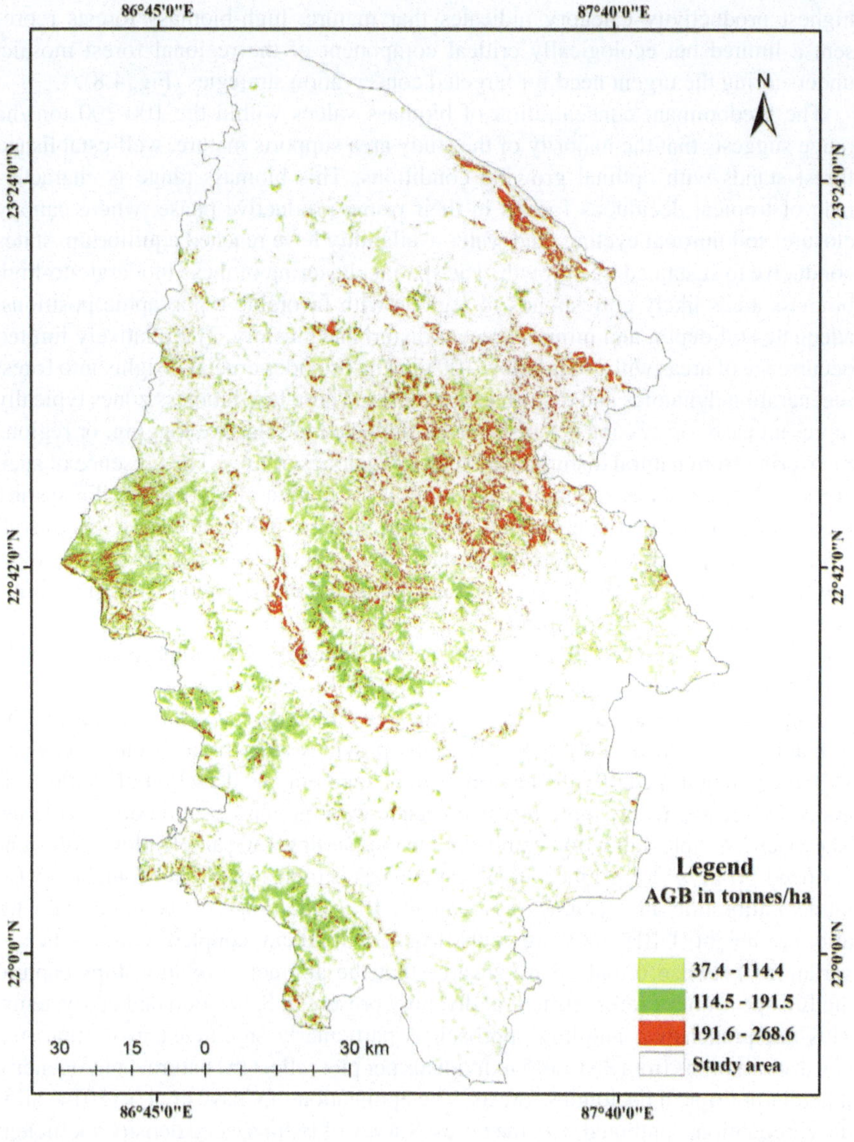

Fig. 4.8 Predicted AGB map of the study area

half (47.58%) of the landscape characterized by relatively modest biomass accumulation (37.4-114.4 tons/ha). This pattern suggests extensive areas of degraded or early successional forests, likely reflecting historical anthropogenic pressures and natural disturbance regimes. The progressive reduction in areal extent with increasing biomass classes from 32.65% in the intermediate range to merely 19.77% in the highest productivity category indicates that mature, high-biomass forests represent a limited but ecologically critical component of the regional forest mosaic, underscoring the urgent need for targeted conservation strategies (Fig. 4.8).

The predominant concentration of biomass values within the 100-200 tons/ha range suggests that the majority of the study area supports mature, well-established forest stands with optimal growing conditions. This biomass range is characteristic of tropical deciduous forests in their prime productive phase, where canopy closure, soil nutrient cycling, and water availability have reached equilibrium states conducive to sustained tree growth. The spatial clustering of these moderate-to-high biomass areas likely corresponds to regions with favorable topographic positions, adequate soil depth, and minimal recent disturbance history. The relatively limited occurrence of areas with AGB below 100 tons/ha provides crucial insights into forest regeneration dynamics and disturbance patterns. These low-biomass zones typically represent early successional forests, areas subjected to selective logging, or regions recovering from natural disturbances such as cyclones or fires. The presence of such areas within the landscape matrix is essential for maintaining forest resilience and biodiversity, as they provide habitat for pioneer species and contribute to the overall structural complexity of the forest ecosystem. Furthermore, these zones serve as indicators of forest management intensity and can guide conservation strategies aimed at promoting natural regeneration.

The integration of Sentinel-2 multispectral imagery with comprehensive field validation demonstrates the evolution of forest biomass assessment from traditional ground-based approaches to sophisticated remote sensing methodologies. The utilization of 10-m resolution VNIR bands provides an optimal balance between spatial detail and spectral information content, enabling the detection of subtle variations in forest structure that directly correlate with biomass accumulation patterns. The extensive field validation campaign, encompassing 130 sample plots with standardized 30 m × 30 m dimensions, establishes a robust empirical foundation for model calibration and accuracy assessment. The wide range of recorded diameter at breast height (DBH) measurements, from 35 to 80 cm, coupled with tree height variations between 6 and 15 m, ensures that the allometric relationships capture the full spectrum of forest structural diversity present in Sal-dominated ecosystems. This comprehensive sampling approach is particularly significant given that tree density variations from 33 to 145 individuals per plot reflect the natural heterogeneity inherent in tropical deciduous forests. The application of Chave et al.'s (2014) allometric equation, combined with the Forest Survey of India's wood density coefficient for Sal (0.73 gm/cm^3), represents a scientifically rigorous approach to biomass estimation that accounts for species-specific growth patterns and wood characteristics. This methodological choice ensures comparability with other regional studies while maintaining accuracy in biomass calculations specific to the dominant tree species in the study area.

The carbon stock mapping reveals patterns that extend beyond simple biomass quantification, providing critical insights into the region's role in global carbon cycling and climate regulation. The spatial congruence between AGB and carbon stock distributions underscores the direct relationship between forest structural complexity and carbon storage capacity, with implications for both local ecosystem services and broader climate mitigation strategies. The identification of 655 km^2 of forest area represents a substantial carbon reservoir that contributes significantly to regional carbon budgets. The average biomass value of 133.51 tons/ha, when converted to carbon stock estimates, indicates that these forests serve as important carbon sinks, actively sequestering atmospheric CO_2 through photosynthetic processes and storing it in woody biomass (Fig. 4.9). The carbon stock distribution across 4043 km^2 exhibits a distinctive bimodal pattern, with low-carbon areas (<25 tons/ha) dominating 44.34% of the landscape, while high-carbon zones (>95 tons/ha) constitute 33.46% of the region. The relatively narrow intermediate class (25-60 tons/ha) occupying only 22.19% suggests limited transitional forest conditions, indicating rapid ecological shifts between degraded and mature forest states (Fig. 4.9). This polarized distribution pattern reflects pronounced landscape heterogeneity, likely driven by varying anthropogenic pressures and natural regeneration processes, creating a mosaic of carbon-depleted and carbon-rich forest patches with significant implications for regional climate mitigation strategies. This carbon storage function becomes increasingly critical in the context of global climate change mitigation efforts, particularly given the rapid deforestation rates observed in many tropical regions. The spatial heterogeneity in carbon stock distribution also reveals important ecological gradients that influence long-term carbon stability. Areas with higher biomass concentrations likely represent carbon storage hotspots that warrant priority conservation attention, while regions with intermediate biomass levels may offer the greatest potential for carbon sequestration enhancement through appropriate silvicultural interventions.

The observed AGB range aligns remarkably well with established benchmarks for tropical deciduous forests, lending credibility to the methodological approach and supporting the broader applicability of the findings. The concordance with studies from similar ecological zones, including Dudhwa National Park and other Sal-dominated landscapes, suggests that the biomass patterns observed in Jangal Mahal are representative of broader regional trends in tropical forest productivity. The statistical performance metrics, while showing moderate correlation coefficients, reflect the inherent challenges in predicting forest biomass across heterogeneous landscapes using remote sensing approaches. The pseudo-R^2 value of 0.48, though lower than some comparable studies, demonstrates acceptable predictive capability when considering the complex topographic and ecological variability characteristic of the study region. The relatively low RMSE of 25.36% indicates that the model maintains reasonable accuracy across the full range of biomass values, making it suitable for landscape-scale forest management and conservation planning applications.

The detailed biomass and carbon stock mapping provides essential baseline information for developing evidence-based forest management strategies that balance conservation objectives with sustainable resource utilization. The identification of high-biomass areas as mature forest stands suggests these regions should receive priority protection status, while low-biomass areas may benefit from active restoration interventions designed to accelerate successional processes and enhance carbon sequestration potential. The comprehensive spatial coverage achieved through

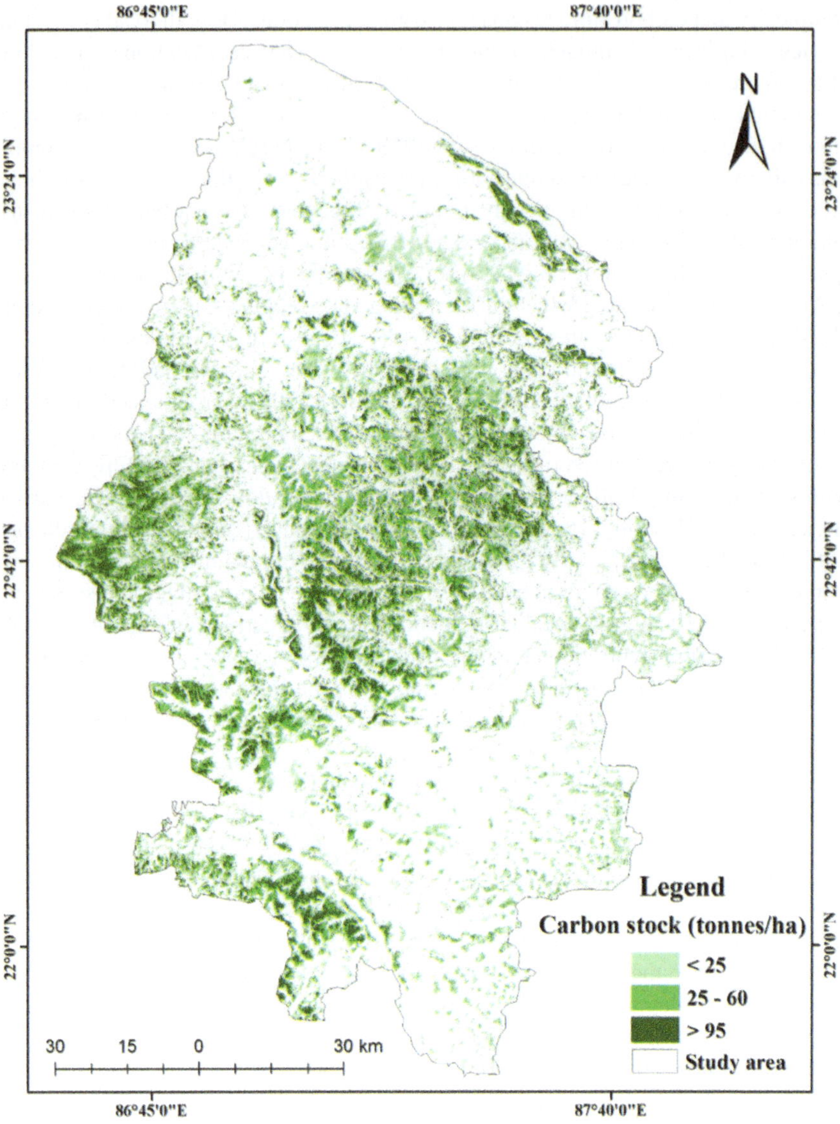

Fig. 4.9 Predicted carbon stock map of the study area

remote sensing approaches offers significant advantages over traditional forest inventory methods, enabling forest managers to develop targeted interventions based on actual biomass distribution patterns rather than generalized estimates. This precision in biomass mapping supports adaptive management strategies that can respond to changing environmental conditions and emerging conservation priorities while maintaining the long-term ecological integrity of these valuable forest ecosystems.

Variation of Above-Ground Biomass in Different Forest Cover Zones

The biomass distribution across forest cover types in the Jangal Mahal region demonstrates a clear ecological hierarchy that underscores the irreplaceable productivity value of indigenous forest systems. Native Sal forests achieve the highest mean AGB of 154.36 tons/ha with substantial structural variation (±17.82), reflecting the complex architectural diversity and optimized resource utilization characteristic of climax forest communities (Table 4.3). This high variability indicates diverse microhabitats, multiple canopy layers, and varied age structures that collectively maximize carbon accumulation potential. The pronounced productivity gaps reveal significant ecological trade-offs associated with plantation forestry. Acacia plantations achieve only 67.5% of Sal Forest biomass (104.26 tons/ha) despite nitrogen-fixing capabilities, while their limited structural variation (±8.81) suggests homogeneous stand conditions that constrain long-term productivity development. Eucalyptus plantations perform even more poorly, reaching merely 57% of native forest productivity (88.12 tons/ha) with minimal structural complexity (±7.45), indicating monocultural constraints and potential allelopathic effects that suppress ecosystem development.

Mixed forests occupy a strategic intermediate position (129.41 tons/ha), demonstrating that species diversification can partially mitigate plantation productivity losses while enhancing ecological complexity. Their moderate structural variation (±12.22) suggests retained ecosystem heterogeneity compared to pure plantations. These findings illuminate fundamental principles governing forest carbon sequestration. The superior performance of Sal forests reflects integrated ecological processes mycorrhizal associations, optimized nutrient cycling, and co-evolved species interactions that cannot be replicated through simplified plantation approaches. The data emphasize that sustainable forest management must prioritize indigenous forest conservation while exploring diversification strategies to enhance plantation productivity and ecosystem resilience.

Table 4.3 AGB among various forest types

Forest type	AGB (Tons/ha)			
	Maximum	Minimum	Mean	SD
Sal Forest	210	138	154.36	17.82
Planted Acacia Forest replacing of Sal	123	102	104.26	8.81
Planted Eucalyptus Forest replacing of Sal	93	68	88.12	7.45
Mixed Forest	138	119	129.41	12.22

4.4.3 Conclusion

This comprehensive assessment of forest biomass and carbon stock dynamics in the Jangal Mahal region establishes critical baseline data that illuminates both the ecological potential and conservation imperatives of West Bengal's tropical deciduous forest ecosystems. The integration of advanced remote sensing methodologies with extensive field validation demonstrates the capability of Sentinel-2 imagery combined with Random Forest algorithms to achieve reliable biomass estimation across heterogeneous forest landscapes, with the weighted extraction approach yielding substantially superior performance ($R^2 = 0.48$, RMSE = 25.36%) compared to conventional centroid-based methods. The pronounced spatial heterogeneity in biomass distribution, ranging from 37.4 to 268.6 tons/ha across 2709 km^2, reflects complex ecological gradients shaped by anthropogenic pressures, natural disturbance regimes, and forest successional dynamics. The bimodal carbon stock distribution pattern, characterized by 44.34% low-carbon areas and 33.46% high-carbon zones, reveals a polarized landscape requiring differentiated management strategies. The dominance of native Sal forests in productivity metrics achieving 48% higher biomass than Acacia plantations and 75% higher than Eucalyptus monocultures provides compelling evidence for prioritizing indigenous species conservation over exotic plantation development.

The research findings carry profound implications for regional climate mitigation strategies, given that these forests represent substantial carbon reservoirs contributing to atmospheric CO_2 regulation. The identification of 655 km^2 of forest area with an average biomass of 133.51 tons/ha establishes the Jangal Mahal region as a significant carbon sink requiring targeted conservation interventions. The superior structural complexity and productivity of mixed forests compared to monocultures suggest that diversified silvicultural approaches can enhance both ecosystem services and biodiversity conservation outcomes. These results provide an essential scientific foundation for implementing evidence-based forest management protocols, establishing monitoring frameworks for REDD+ mechanisms, and developing adaptive strategies that balance conservation objectives with sustainable resource utilization, ultimately supporting long-term ecosystem integrity and climate resilience in this ecologically vital region of eastern India.

References

Allen CD, Macalady AK, Chenchouni H, Bachelet D, McDowell N, Vennetier M, Cobb N et al (2010) A global overview of drought and heat-induced tree mortality reveals emerging climate change risks for forests. For Ecol Manage 259(4):660–684

Angelsen A (Ed) (2009) Realising REDD+: National strategy and policy options. Cifor

Asner GP, Mascaro J, Anderson C, Knapp DE, Martin RE, Kennedy-Bowdoin T, Bermingham E et al (2013) High-fidelity national carbon mapping for resource management and REDD+. Carbon Balance Manage 8(1):7

Avitabile V, Herold M, Heuvelink GB, Lewis SL, Phillips OL, Asner GP, Willcock S et al (2016) An integrated pan-tropical biomass map using multiple reference datasets. Glob Change Biol 22(4):1406–1420

Baccini A, Walker W, Carvalho L, Farina M, Sulla-Menashe D, Houghton RA (2017) Tropical forests are a net carbon source based on aboveground measurements of gain and loss. Science 358(6360):230–234

Bradshaw CJ, Warkentin IG (2015) Global estimates of boreal forest carbon stocks and flux. Global Planet Change 128:24–30

Baishya R, Barik SK (2011) Estimation of tree biomass, carbon pool and net primary production of an old-growth Pinus kesiya Royle ex. Gordon forest in north-eastern India. Ann Forest Sci 68(4) 727–736.

Breiman L (2001) Random Forests. Mach Learn 45(1):5–32

Cairns MA, Brown S, Helmer EH, Baumgardner GA (1997) Root biomass allocation in the world's upland forests. Oecologia 111(1):1–11

Chanda A, Akhand A (2023) Challenges towards the sustainability and enhancement of the Indian Sundarban mangrove's blue carbon stock. Life 13(8):1787

Chave J, Réjou-Méchain M, Búrquez A, Chidumayo E, Colgan MS, Delitti WB, Vieilledent G et al (2014) Improved allometric models to estimate the aboveground biomass of tropical trees. Glob Change Biol 20(10):3177–3190

Chitale VS, Behera MD, Roy PS (2014) Future of endemic flora of biodiversity hotspots in India. PLoS ONE 9(12):e115264

Das GK (2021) Forests and forestry of West Bengal: survey and analysis. Springer Nature.

Dong L, Du H, Han N, Li X, Zhu DE, Mao F, He S et al (2020) Application of convolutional neural network on lei bamboo above-ground-biomass (AGB) estimation using Worldview-2. Remote Sensing 12(6):958

Duncanson L, Armston J, Disney M, Avitabile V, Barbier N, Calders K, Williams M et al (2019) The importance of consistent global forest aboveground biomass product validation. Surv Geophys 40(4):979–999

Dubayah R, Blair JB, Goetz S, Fatoyinbo L, Hansen M, Healey S, Silva C et al (2020) The global ecosystem dynamics investigation: high-resolution laser ranging of the Earth's forests and topography. Sci Remote Sensing 1:100002

Dutta Roy A, Das SK, Debnath B (2023) A non-destructive approach to develop tree-level allometric equations for estimating aboveground biomass in the forests of Tripura, Northeast India. Tropical Ecol 64(3):532–542

European Space Agency (2015) Sentinel-2 User Handbook. ESA Standard Document; European Space Agency, Paris, France

Friedlingstein P, O'sullivan M, Jones MW, Andrew RM, Hauck J, Olsen A, Zaehle S et al (2020) Global carbon budget 2020. Earth Syst Sci Data Discussions 2020:1–3

Forest Survey of India (2021) India State of Forest Report 2021. Ministry of Environment, Forest and Climate Change, Government of India. https://fsi.nic.in/isfr2021/

Griscom BW, Adams J, Ellis PW, Houghton RA, Lomax G, Miteva DA, Fargione J et al (2017) Natural climate solutions. Proc Natl Acad Sci 114(44):11645–11650

Ghosh A, Schmidt S, Fickert T, Nüsser M (2015) The Indian Sundarban mangrove forests: history, utilization, conservation strategies and local perception. Diversity 7(2):149–169

González-Jaramillo V, Fries A, Bendix J (2019) AGB estimation in a tropical mountain forest (TMF) by means of RGB and multispectral images using an unmannedaerial vehicle (UAV). Remote Sensing 11(12):1413

Gorelick N, Hancher M, Dixon M, Ilyushchenko S, Thau D, Moore R (2017) Google earth engine: planetary-scale geospatial analysis for everyone. Remote Sens Environ 202:18–27

Hansen MC, Potapov PV, Moore R, Hancher M, Turubanova SA, Tyukavina A, Townshend JR, et al (2013) High-resolution global maps of 21st-century forest cover change. Science 342(6160), 850–853

Haralick RM, Shanmugam K, Dinstein IH (2007) Textural features for image classification. IEEE Trans Syst Man Cybern 6:610–621
Hasanuzzaman MD, Adhikary PP, Shit PK (2024) Gully erosion susceptibility mapping and prioritization of gully-dominant sub-watersheds using machine learning algorithms: Evidence from the Silabati River (tropical river, India). Adv Space Res 73(3):1653–1666
Hirata Y, Tabuchi R, Patanaponpaiboon P, Poungparn S, Yoneda R, Fujioka Y (2014) Estimation of aboveground biomass in mangrove forests using high-resolution satellite data. J Res 19(1):34–41
Joshi PK, Roy PS, Singh S, Agrawal S, Yadav D (2006) Vegetation cover mapping in India using multi-temporal IRS Wide Field Sensor (WiFS) data. Remote Sens Environ 103(2):190–202
Jw R (1973) Monitoring vegetation systems in the great plains with ERTS. In: Third NASA earth resources technology satellite symposium, 1973, vol 1, pp 309–317
Karmakar D, Gupta S, Padhy PK (2023) A critical review of different methods of estimation of the above-ground biomass and carbon stocks in India. Ecophysiol Trop Plants 233–252
Kelsey KC, Neff JC (2014) Estimates of aboveground biomass from texture analysis of Landsat imagery. Remote Sensing 6(7):6407–6422
Köhl M, Baldauf T, Plugge D, Krug J (2009) Reduced emissions from deforestation and forest degradation (REDD): a climate change mitigation strategy on a critical track. Carbon Balance Manage 4(1):10
Kothamasi D, Kothamasi S, Bhattacharyya A, Kuhad RC, Babu CR (2006) Arbuscular mycorrhizae and phosphate solubilising bacteria of the rhizosphere of the mangrove ecosystem of Great Nicobar island, India. Biology and Fertility of Soils 42(4):358–361
Kurz WA, Dymond CC, Stinson G, Rampley GJ, Neilson ET, Carroll AL, Safranyik L et al (2008) Mountain pine beetle and forest carbon feedback to climate change. Nature 452(7190):987–990
Li Y, Li C, Li M, Liu Z (2019) Influence of variable selection and forest type on forest aboveground biomass estimation using machine learning algorithms. Forests 10(12):1073
Lorenz K, Lal R (2018) Soil carbon stock. Carbon sequestration in agricultural ecosystems. Springer International Publishing, Cham, pp 39–136
Luyssaert S, Schulze ED, Börner A, Knohl A, Hessenmöller D, Law BE, Grace J et al (2008) Old-growth forests as global carbon sinks. Nature 455(7210):213–215
Martin AR, Doraisami M, Thomas SC (2018) Global patterns in wood carbon concentration across the world's trees and forests. Nat Geosci 11(12):915–920
Matsushita B, Yang W, Chen J, Onda Y, Qiu G (2007) Sensitivity of the enhanced vegetation index (EVI) and normalized difference vegetation index (NDVI) to topographic effects: a case study in high-density cypress forest. Sensors 7(11):2636–2651
Marvin DC, Asner GP, Knapp DE, Anderson CB, Martin RE, Sinca F, Tupayachi R (2014) Amazonian landscapes and the bias in field studies of forest structure and biomass. Proc Natl Acad Sci 111(48):E5224–E5232
Mandal RN, Naskar KR (2008) Diversity and classification of Indian mangroves: a review
Mokany K, Raison RJ, Prokushkin AS (2006) Critical analysis of root: shoot ratios in terrestrial biomes. Glob Change Biol 12(1):84–96
Mukhopadhyay A, Mitra D, Hazra S (eds) (2021) Sundarbans mangrove systems: A geo-informatics approach. CRC Press
Murthy IK, Gupta M, Tomar S, Munsi M, Tiwari R, Hegde GT, Ravindranath NH (2013) Carbon sequestration potential of agroforestry systems in India. J Earth Sci Climate Change 4(1):1–7
Mukherjee P, Ray B, Bhattacharya RN (2017) Status differences in collective action and forest benefits: evidence from joint forest management in India. Environ Dev Sustain 19(5):1831–1854
Nair PR, Kumar BM, Nair VD (2022) Soil organic matter (SOM) and nutrient cycling. An introduction to agroforestry: four decades of scientific developments. Springer International Publishing, Cham, pp 383–411
Nandy S, Kushwaha SPS, Dadhwal VK (2011) Forest degradation assessment in the upper catchment of the river Tons using remote sensing and GIS. Ecol Ind 11(2):509–513

Naidoo L, Van Deventer H, Ramoelo A, Mathieu R, Nondlazi B, Gangat R (2019) Estimating above ground biomass as an indicator of carbon storage in vegetated wetlands of the grassland biome of South Africa. Int J Appl Earth Obs Geoinf 78:118–129

Nunery JS, Keeton WS (2010) Forest carbon storage in the northeastern United States: net effects of harvesting frequency, post-harvest retention, and wood products. For Ecol Manage 259(8):1363–1375

Pan Y, Birdsey RA, Fang J, Houghton R, Kauppi PE, Kurz WA, Hayes D (2011) A large and persistent carbon sink in the world's forests. Science 333(6045), 988–993

Paul S, Das CS (2025) Seasonal variation of sedimentary nutrients and litter production in the mangrove ecosystem of indian sundarban: implications for ecosystem dynamics. Plant Soil 511(1):263–286

Prasad AM, Iverson LR, Liaw A (2006) Newer classification and regression tree techniques: bagging and random forests for ecological prediction. Ecosystems 9(2):181–199

Qasim M, Hubacek K, Termansen M, Fleskens L (2013) Modelling land use change across elevation gradients in district Swat, Pakistan. Reg Environ Change 13(3):567–581

Ray R, Ganguly D, Chowdhury C, Dey M, Das S, Dutta MK, Jana TK et al (2011) Carbon sequestration and annual increase of carbon stock in a mangrove forest. Atmos Environ 45(28):5016–5024

Rajput S, Mohanta A, Banerjee B, Das J, Mishra S, Shankar Kumar H (2025) Harnessing GIS, remote sensing, and machine learning for sustainable management and carbon sequestration of non-timber forest products in Gujarat, India. Agroforestry Systems 99(5):92

Rodríguez-Veiga P, Saatchi S, Tansey K, Balzter H (2016) Magnitude, spatial distribution and uncertainty of forest biomass stocks in Mexico. Remote Sens Environ 183:265–281

Reddy CS, Sreelekshmi S, Jha CS, Dadhwal VK (2013) National assessment of forest fragmentation in India: landscape indices as measures of the effects of fragmentation and forest cover change. Ecol Eng 60:453–464

Rejou-Mechain M, Muller-Landau HC, Detto M, Thomas SC, Le Toan T, Saatchi SS, Chave J et al (2014) Local spatial structure of forest biomass and its consequences for remote sensing of carbon stocks. Biogeosciences 11(23):6827–6840

Réjou-Méchain M, Tanguy A, Piponiot C, Chave J, Hérault B (2017) Biomass: an r package for estimating above-ground biomass and its uncertainty in tropical forests. Methods Ecol Evol 8(9):1163–1167

Roy S, Mudi S, Das P, Ghosh S, Shit PK, Bhunia GS, Kim J (2020) Estimating above ground biomass (AGB) and tree density using sentinel-1 data. Spatial modeling in forest resources management: rural livelihood and sustainable development. Springer International Publishing, Cham, pp 259–280

Saatchi SS, Harris NL, Brown S, Lefsky M, Mitchard ET, Salas W, Morel A et al (2011) Benchmark map of forest carbon stocks in tropical regions across three continents. Proc Natl Acad Sci 108(24):9899–9904

Saha D, Sundriyal RC (2012) Utilization of non-timber forest products in humid tropics: Implications for management and livelihood. Forest Policy Econ 14(1):28–40

Santoro M, Beaudoin A, Beer C, Cartus O, Fransson JE, Hall RJ, Wegmüller U et al (2015) Forest growing stock volume of the northern hemisphere: spatially explicit estimates for 2010 derived from Envisat ASAR. Remote Sens Environ 168:316–334

Santoro M, Cartus O, Carvalhais N, Rozendaal DM, Avitabile V, Araza A, Willcock S et al (2021) The global forest above-ground biomass pool for 2010 estimated from high-resolution satellite observations. Earth Syst Sci Data 13(8):3927–3950

Sharma CM, Baduni NP, Gairola S, Ghildiyal SK, Suyal S (2010) Tree diversity and carbon stocks of some major forest types of Garhwal Himalaya, India. For Ecol Manage 260(12):2170–2179

Sen S, Zipper CE, Wynne RH, Donovan PF (2012) Identifying revegetated mines as disturbance/recovery trajectories using an interannual Landsat chronosequence. Photogramm Eng Remote Sens 78(3):223–235

Shit PK, Pourghasemi HR, Das P, Bhunia GS (2020) Spatial modeling in forest resources management. Springer, Switzerland

Stephenson NL, Das AJ, Condit R, Russo SE, Baker PJ, Beckman NG, Zavala MA et al (2014) Rate of tree carbon accumulation increases continuously with tree size. Nature 507(7490):90–93

Seidl R, Thom D, Kautz M, Martin-Benito D, Peltoniemi M, Vacchiano G, Reyer CP et al (2017) Forest disturbances under climate change. Nat Clim Chang 7(6):395–402

Schuur EA, McGuire AD, Schädel C, Grosse G, Harden JW, Hayes DJ, Vonk JE et al (2015) Climate change and the permafrost carbon feedback. Nature 520(7546):171–179

Silva CA, Hudak AT, Klauberg C, Vierling LA, Gonzalez-Benecke C, de Padua Chaves Carvalho S, Cardil A (2017) Combined effect of pulse density and grid cell size on predicting and mapping aboveground carbon in fast-growing Eucalyptus forest plantation using airborne LiDAR data. Carbon Balance Manage 12(1):13

Sibanda M, Mutanga O, Rouget M (2015) Examining the potential of Sentinel-2 MSI spectral resolution in quantifying above ground biomass across different fertilizer treatments. ISPRS J Photogramm Remote Sens 110:55–65

Thapa B, Lovell S, Wilson J (2023) Remote sensing and machine learning applications for aboveground biomass estimation in agroforestry systems: a review. Agrofor Syst 97(6):1097–1111

Thomas SC, Martin AR (2012) Carbon content of tree tissues: a synthesis. Forests 3(2):332–352

Tolangay D, Moktan S (2024) An insight into woody species in sub-temperate forest ecosystem in Darjeeling Himalaya, India. J Trop Forest Sci 36(1):51–67

Tucker CJ (1979) Red and photographic infrared linear combinations for monitoring vegetation. Remote Sens Environ 8(2):127–150

Usuga JCL, Toro JAR, Alzate MVR, Tapias ÁDJL (2010) Estimation of biomass and carbon stocks in plants, soil and forest floor in different tropical forests. For Ecol Manage 260(10):1906–1913

Xu L, Saatchi SS, Yang Y, Yu Y, Pongratz J, Bloom AA, Schimel D, et al (2021) Changes in global terrestrial live biomass over the 21st century. Sci Adv 7(27):eabe9829

Chapter 5
Urban Expansion and Forest Footprint: Emerging Conflicts

Abstract This chapter investigates the escalating spatial conflicts between urban expansion and forest ecosystems in Siliguri Town and its surrounding areas within West Bengal's Eastern Himalayan biodiversity hotspot. Employing multi-temporal satellite imagery analysis using Landsat-5 TM (1990) and Landsat-8 OLI (2025) data through supervised Maximum Likelihood Classification, the study quantifies dramatic land-use/land-cover transformations over a 35-year period within a 20-km radius. Results reveal unprecedented urban expansion from 64.17 km^2 (5.11%) to 226.68 km^2 (18.05%), representing a 253% increase that fundamentally altered regional landscape dynamics. Concurrently, vegetation covers experienced substantial decline from 490.14 km^2 (39.03%) to 433.85 km^2 (34.55%), with 56.29 km^2 lost primarily through conversion to agricultural land (76.42 km^2) and built-up areas (25.45 km^2). This rapid urbanization has fragmented critical ecological corridors connecting Mahananda Wildlife Sanctuary, Jaldapara National Park, and Gorumara National Park, intensifying human-elephant conflicts with over 150 documented incidents annually. The study demonstrates how Siliguri's strategic position as Northeast India's commercial gateway has accelerated deforestation through infrastructure development, illegal settlements, and industrial expansion, particularly impacting the ecologically sensitive Terai-Duar savanna ecosystem. Post-classification comparison and transition intensity analysis reveal that urban sprawl disproportionately affects forest buffer zones and wildlife migration routes, compromising ecosystem service provision including flood regulation, carbon sequestration, and groundwater recharge. The 2017 monsoon floods exemplified these vulnerabilities, where deforested watersheds exacerbated urban flooding affecting 200,000 residents. Findings underscore the urgent necessity for integrated landscape governance frameworks bridging institutional fragmentation between municipal authorities and forest departments, implementing green infrastructure planning, establishing urban growth boundaries, and restoring ecological corridors to reconcile development imperatives with biodiversity conservation in this geopolitically sensitive border region.

Keywords Urban expansion · Land-use/land-cover change · Remote sensing and GIS · Eastern Himalayan biodiversity hotspot · Human-wildlife conflict

P. K. Shit and M. Hasanuzzaman, *Forest Landscape Dynamics in West Bengal*, Advances in Asian Human-Environmental Research,
https://doi.org/10.1007/978-3-032-23700-2_5

5.1 Introduction

The accelerating pace of urban expansion has generated unprecedented conflicts between growing metropolitan areas and adjacent forest ecosystems, creating complex socio-ecological tensions that challenge contemporary approaches to sustainable land-use planning (Wang et al. 2023). As global urban populations approach 68% by 2050, the spatial footprint of cities expands exponentially, directly encroaching upon forest landscapes that provide critical ecosystem services including carbon sequestration, biodiversity conservation, and hydrological regulation (UN Desa 2018). This urban sprawl phenomenon has emerged as a primary driver of forest fragmentation, with studies indicating that urban growth accounts for approximately 15-20% of annual global forest loss, particularly concentrated in rapidly developing regions of Southeast Asia, Latin America, and sub-Saharan Africa (Hansen et al. 2013). The ecological dimensions of urban-forest conflicts manifest through direct habitat conversion and indirect edge effects that penetrate up to 500 m into forest interiors, altering microclimate conditions and species composition patterns (Laurance et al. 2007). Remote sensing analyses reveal that forest conversion rates to urban uses have increased by 300% since 1990, with particularly severe impacts occurring in biodiversity hotspots such as the Western Ghats and Brazilian Cerrado (Curtis et al. 2018). These rapid transformations create cascading effects on ecosystem functionality, disrupting wildlife corridors and reducing the capacity of forest fragments to maintain viable populations of native species.

Economic conflicts arise from fundamental market failures that fail to internalize the estimated $150 trillion annual value of global forest ecosystem services, creating perverse incentives favoring immediate urban development returns over long-term ecological benefits (Costanza et al. 2014). This economic asymmetry becomes particularly pronounced in developing regions where weak institutional frameworks enable urban expansion into protected forest areas without adequate environmental impact assessment or compensation mechanisms. Social dimensions of these conflicts center on environmental justice concerns as urban expansion displaces forest-dependent communities and disrupts traditional land tenure systems (Larson 2010). Research in Indonesia's Kalimantan region documents how urban development and associated infrastructure have displaced over 50,000 indigenous Dayak people, severing cultural connections to ancestral forest landscapes while providing minimal economic compensation (Obidzinski et al. 2012). Similar patterns of displacement occur globally, creating what geographers' term "ecological marginalization" where vulnerable populations bear disproportionate costs of urban-driven deforestation.

Climate change amplifies urban-forest conflicts through feedback mechanisms that simultaneously increase urban vulnerability to extreme weather events while degrading the forest ecosystems upon which cities depend for climate adaptation services (Barona 2015). Urban heat island effects extending into adjacent forests modify local precipitation patterns and temperature regimes, while cities increasingly require forest-based solutions for flood control and temperature regulation (Grimm et al. 2008). This creates paradoxical situations where urban expansion destroys the ecological infrastructure necessary for urban climate resilience, highlighting the urgent need for integrated landscape management approaches that recognize the interdependence of urban and forest systems.

5.2 Urban Expansion and Forest Footprint: Emerging Conflicts in West Bengal

West Bengal presents a critical case study of urban-forest conflicts in South Asia, where rapid metropolitan expansion has generated intense pressure on the state's remaining forest ecosystems (Shit et al. 2021). The Kolkata Metropolitan Area (KMA), India's third-largest urban agglomeration, has experienced unprecedented sprawl extending into the ecologically sensitive Eastern Ghats and Sundarbans periphery, resulting in a 35% reduction in forest cover since 1990 (Nandy et al. 2019). This expansion has particularly impacted the Mahananda Wildlife Sanctuary and Buxa Tiger Reserve corridors, where urban infrastructure development has fragmented critical elephant migration routes and tiger habitats. The state's industrial belt extending from Kolkata to Durgapur has accelerated deforestation rates, with coal mining and steel production creating additional pressure on the remaining 11.8% forest cover, significantly below the national average of 21.67% (Forest Survey of India 2021). Urban encroachment into the Sundarbans buffer zones presents unique challenges, as rising sea levels and increasing cyclone intensity compound anthropogenic pressures, threatening both mangrove ecosystems and urban coastal settlements.

Social conflicts have intensified as tribal communities, particularly Santhals and Oraons, face displacement from traditional forest lands converted for urban expansion and industrial projects (Iyer 2020). The Singur and Nandigram controversies exemplify these tensions, where agricultural and forest land acquisition for urban-industrial development generated widespread resistance and political upheaval (Banerjee-Guha 2008). Economic pressures drive continued encroachment, as urban land values in peri-urban forest areas have increased by 400% between 2010 and 2020, creating speculative pressures that overwhelm forest protection mechanisms. Climate change exacerbates these conflicts through increased frequency of extreme weather events, with urban areas dependent on forest ecosystems for flood control and temperature regulation. The 2009 Cyclone Aila demonstrated this interdependence, as mangrove deforestation amplified storm surge impacts on Kolkata's eastern suburbs (Mukhopadhyay et al. 2018). Contemporary planning approaches emphasize green corridor development and urban-forest restoration, yet implementation faces challenges from institutional fragmentation between municipal corporations and forest departments, highlighting the need for integrated landscape management approaches that address West Bengal's unique geographical and socio-political context (Fig. 5.1).

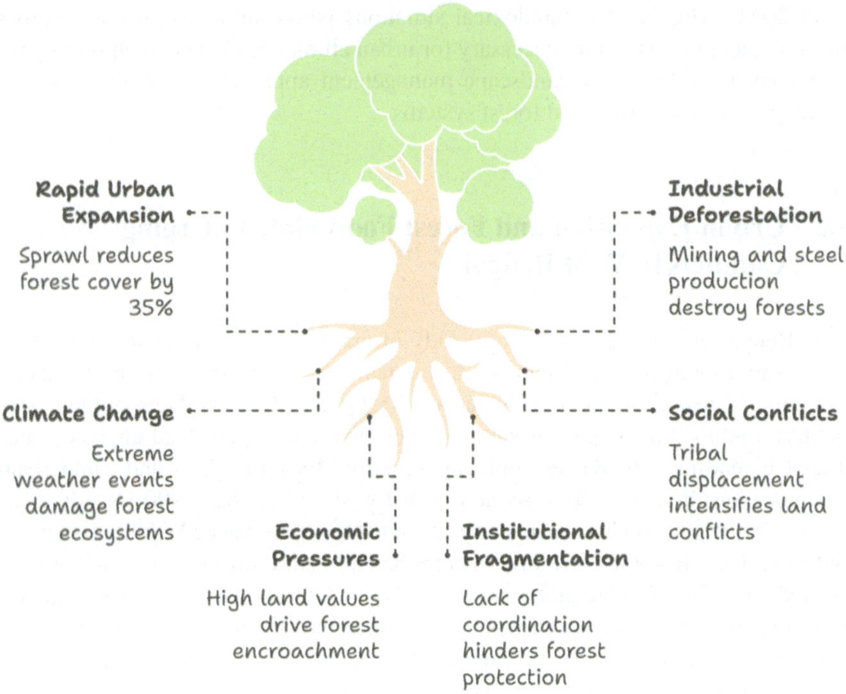

Fig. 5.1 Symmetrical diagram of urban-forest conflicts in West Bengal

5.3 Urban Expansion and Forest Footprint: Emerging Conflicts—GIS Calculation Tools and Techniques

Geographic Information Systems (GIS) have emerged as indispensable tools for quantifying and analyzing urban-forest conflicts, providing sophisticated methodologies for measuring spatial patterns, temporal dynamics, and predictive modeling of land-cover transitions (Fig. 5.2). Contemporary GIS applications integrate multi-temporal satellite imagery analysis using platforms such as Google Earth Engine and ArcGIS Pro to calculate forest loss rates, urban expansion vectors, and fragmentation indices with unprecedented precision (Gorelick et al. 2017). The Normalized Difference Vegetation Index (NDVI) and Enhanced Vegetation Index (EVI) derived from Landsat and Sentinel-2 imagery enable researchers to quantify vegetation health and forest canopy changes over time, while land-cover classification algorithms including random forest and support vector machine approaches achieve accuracy rates exceeding 90% in distinguishing urban from forest land uses (Pelletier et al. (2016).

Spatial metrics analysis through tools like FRAGSTATS calculates critical landscape ecology indices including patch density, edge density, and connectivity metrics that quantify forest fragmentation patterns resulting from urban encroachment

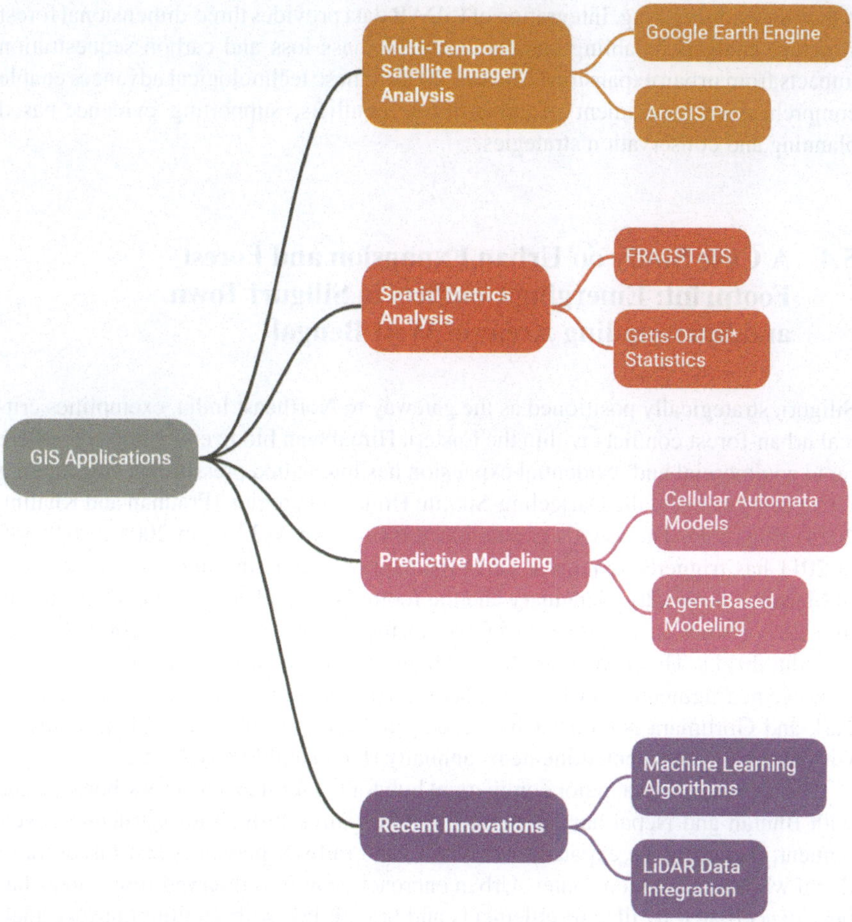

Fig. 5.2 Flow diagram of GIS applications in urban-forest conflict analysis

(McGarigal et al. 2002). Buffer analysis and proximity modeling techniques assess urban pressure zones around forest boundaries, while hotspot analysis using Getis-Ord Gi* statistics identifies statistically significant clusters of forest loss (Rossi and Becker 2019). Advanced cellular automata models and agent-based modeling frameworks simulate future urban expansion scenarios under different policy conditions, incorporating variables such as topography, road networks, and zoning regulations to predict potential forest conversion areas.

Recent innovations include machine learning algorithms for change detection analysis, utilizing convolutional neural networks to automatically identify urban-forest boundary shifts with temporal accuracy of less than 1 year (Zhu et al. 2019). Cloud-based processing platforms enable real-time monitoring of deforestation events, while web-based GIS dashboards facilitate stakeholder engagement and policy decision-making. Integration of LiDAR data provides three-dimensional forest structure analysis, enabling calculation of biomass loss and carbon sequestration impacts from urban expansion (Ali et al. 2025). These technological advances enable comprehensive assessment of urban-forest conflicts, supporting evidence-based planning and conservation strategies.

5.4 A Case Study on Urban Expansion and Forest Footprint: Emerging Conflicts in Siliguri Town and Surrounding Areas of West Bengal

Siliguri, strategically positioned as the gateway to Northeast India, exemplifies critical urban-forest conflicts within the Eastern Himalayan biodiversity hotspot, where rapid commercial and residential expansion has intensified pressure on surrounding forest ecosystems of the Darjeeling-Sikkim Himalayan region (Pradhan and Khaling 2024) (Fig. 5.3). The town's population growth from 472,374 in 2001 to 648,389 in 2011 has triggered extensive peri-urban sprawl extending into the buffer zones of Mahananda Wildlife Sanctuary and the foothills of the Darjeeling Hills, resulting in a 28% reduction in forest cover within the 20-km radius since 2000 (Census of India 2011). This expansion has particularly impacted the Terai-Duar savanna ecosystem, fragmenting critical elephant corridors connecting Jaldapara National Park and Gorumara National Park, leading to increased human-elephant conflicts with over 150 documented incidents annually (Panja and Mistri 2025).

The town's role as a major commercial hub for tea, timber, and cross-border trade with Bhutan and Nepal has accelerated deforestation through infrastructure development, including the expansion of NH-31 and railway networks that bisect traditional wildlife migration routes. Urban encroachment into reserved forest areas has been exacerbated by illegal settlements and brick kilns, with satellite imagery analysis revealing a 15% annual increase in built-up area expansion into forest margins between 2015 and 2020 (Roy et al. 2016). The Mahananda River floodplains, crucial for maintaining ecological connectivity, have experienced significant alteration due to urban waste discharge and construction activities.

Climate vulnerabilities intensify these conflicts as Siliguri faces increasing monsoon variability and flash flooding events, with the town's dependency on surrounding forest watersheds for flood control and groundwater recharge. The 2017 monsoon floods highlighted this interdependence when deforested catchments contributed to unprecedented urban flooding, affecting over 200,000 residents (Polgar and Jaafar 2018). Contemporary conservation efforts include the proposed Siliguri-Jalpaiguri forest corridor restoration project, yet implementation faces challenges from competing land-use demands and weak enforcement of forest protection regulations in this geopolitically sensitive border region.

Fig. 5.3 Representative field photographs of the sal-dominated forest landscape in the study area. **a** Mature sal tree within a mixed forest stand showing undergrowth and surrounding terrain. **b** Tall sal trees forming a relatively open canopy structure. **c** Moderately dense forest patch with leaf-litter accumulation on the forest floor. **d** Mixed woodland exhibiting variable tree spacing and understory vegetation. **e** Managed forest section with visible fencing and thick ground litter. **f** Sloping forest terrain with scattered trees and natural regeneration

5.4.1 Materials and Methods

This study utilizes multi-temporal satellite imagery from the Landsat program to analyze land-use/land-cover (LULC) changes over a 35-year period. Specifically, Landsat-5 Thematic Mapper (TM) data from 1990 and Landsat-8 Operational Land Imager (OLI) data from 2025 were acquired from the United States Geological

Survey (USGS) Earth Explorer platform (https://earthexplorer.usgs.gov/). These datasets were selected for their moderate spatial resolution (30 m) and proven effectiveness in LULC classification across various temporal and geographical contexts.

To ensure geometric consistency and spatial accuracy, all images were co-registered using a first-order polynomial transformation model, with a root mean square error (RMSE) of less than 0.5 pixels. A minimum of five well-distributed ground control points (GCPs) were used for each scene, derived from topographic maps and Google Earth references. Preprocessing steps, including radiometric calibration, atmospheric correction, and mosaicking, were carried out to enhance the quality of the data for subsequent analysis.

The study area's LULC was classified into five major categories: (i) waterbodies, (ii) vegetation, (iii) agricultural land, (iv) built-up area, and (v) other areas. A supervised classification approach was employed using the Maximum Likelihood Classification (MLC) algorithm, which is widely regarded as one of the most robust and statistically sound classification techniques (Du et al. 2013; Zhang et al. 2020). The classification process was executed in the ArcGIS 10.x environment, and training samples for each class were generated through field validation and visual interpretation of high-resolution Google Earth imagery.

To assess spatio-temporal dynamics, a post-classification comparison method was applied to detect LULC changes between 1990 and 2025. This approach involved the generation of transition matrices, including "from-to" pixel change matrices, area conversion matrices, and percentage change matrices (Koko et al. 2022). These tools enabled the identification of specific land transformations particularly the conversion of vegetation and agricultural land into built-up areas. Further, transition intensity analysis was performed using the TerrSet software (formerly IDRISI), focusing on the quantification of vegetation loss and urban sprawl. The Land Change Modeler (LCM) module within TerrSet provided a pixel-by-pixel breakdown of land-cover transitions, offering insights into the spatial patterns and rates of urban expansion and deforestation (Atef et al. 2024). This integrative approach supports a nuanced understanding of anthropogenic pressures on forest ecosystems in the context of rapid urban development.

5.4.2 Results and Discussion

The LULC analysis of Siliguri Town and its surrounding areas within a 20 km radius reveals significant transformations over the 35-year period from 1990 to 2025 (Fig. 5.4 and Table 5.1). The classification maps demonstrate pronounced changes in land-use patterns, reflecting the rapid urbanization characteristic of this strategically located urban center in West Bengal.

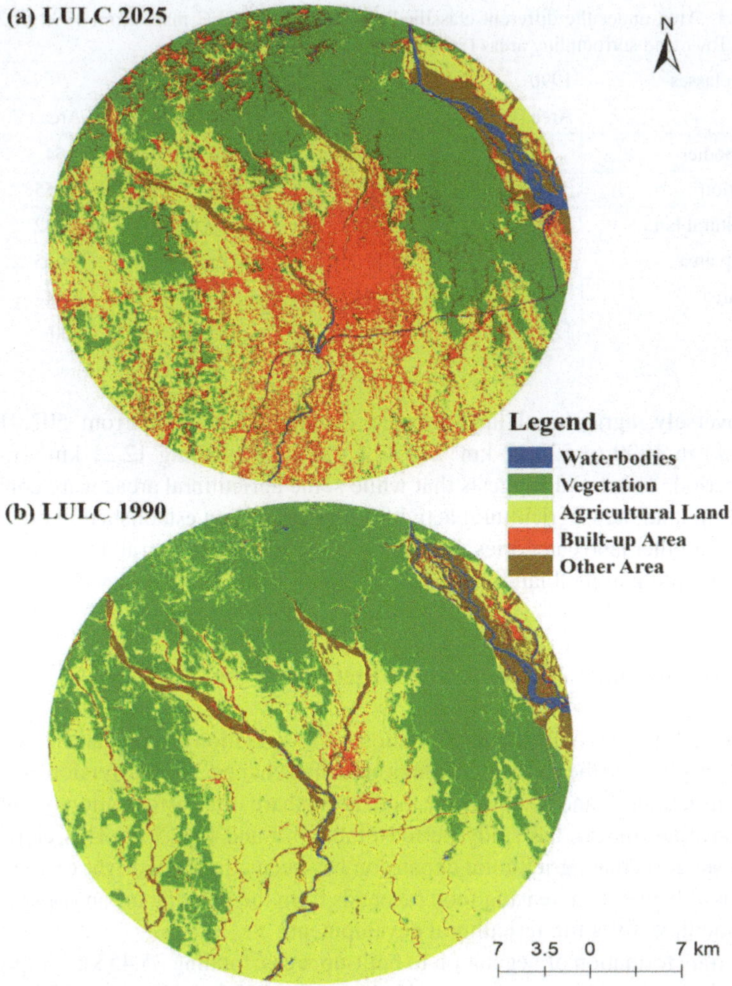

Fig. 5.4 LULC classification maps of the years 1990 and 2025 at Siliguri Town and surrounding areas (20 km) of West Bengal

Vegetation Cover Decline and Agricultural Land Stability

The most notable change observed in the study area is the substantial decline in vegetation cover, which decreased from 490.14 km^2 (39.03%) in 1990 to 433.85 km^2 (34.55%) in 2025, representing a net loss of 56.29 km^2 or approximately 4.48% of the total study area (Table 5.1). This reduction in vegetation cover highlights the mounting pressure on natural ecosystems due to anthropogenic activities and urban expansion processes.

Table 5.1 Area under the different classified classes of the LULC maps from 1990 to 2025 at Siliguri Town and surrounding areas (20 km) of West Bengal

LULC classes	1990		2025	
	Area (km^2)	Area (%)	Area (km^2)	Area (%)
Water bodies	17.25	1.37	19.36	1.54
Vegetation	490.14	39.03	433.85	34.55
Agricultural land	507.91	40.45	520.12	41.42
Built-up area	64.17	5.11	226.68	18.05
Other area	176.33	14.04	55.79	4.44
Total	1255.8	100	1255.8	100

Conversely, agricultural land showed a marginal increase from 507.91 km^2 (40.45%) in 1990 to 520.12 km^2 (41.42%) in 2025, gaining 12.21 km^2 over the study period. This trend suggests that while some agricultural areas were converted to built-up land, new agricultural activities may have been established in previously forested or other land categories, potentially indicating agricultural intensification or expansion into marginal lands.

Land Transformation Patterns from Vegetation

The detailed analysis of land transformation from vegetation to other land uses reveals critical insights into the pathways of environmental change. The conversion of vegetation to agricultural land accounts for the largest share of transformation, with 76.42 km^2 converted during the study period (Table 5.2 and Fig. 5.5). This conversion pattern suggests that agricultural expansion has been a primary driver of vegetation loss, possibly due to increasing food demands from the growing urban population or economic incentives for agricultural development.

The transformation of vegetation to built-up areas, totaling 25.45 km^2, represents direct urban encroachment into natural habitats (Table 5.2 and Fig. 5.5). This conversion pattern is particularly concerning from an ecological perspective, as it represents permanent loss of vegetative cover and associated ecosystem services. The spatial distribution of this transformation, as illustrated in the maps, likely follows transportation corridors and areas of economic development, creating fragmented forest patches and altered landscape connectivity.

Table 5.2 Land transformation from vegetation to other land uses at Siliguri Town and surrounding areas (20 km) of West Bengal

Land transformation	Area in km^2
Vegetation to waterbody	0.89
Vegetation to agricultural land	76.42
Vegetation to built-up area	25.45
Vegetation to other area	1.3

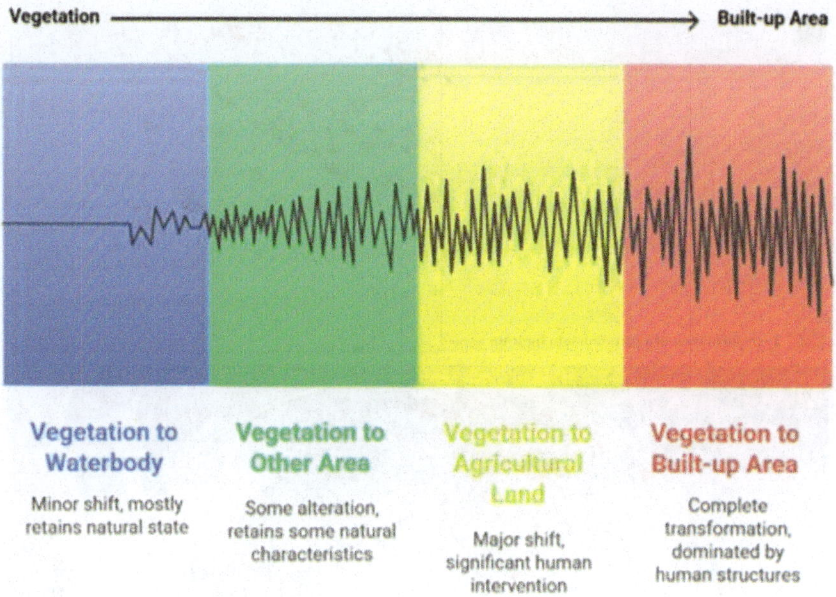

Fig. 5.5 Land transformation from vegetation to other land uses at Siliguri Town and surrounding areas (20 km) of West Bengal

The reduction in "Other Area" category from 176.33 km^2 (14.04%) to 55.79 km^2 (4.44%) indicates significant land-use intensification, with previously unused or marginal lands being converted to more intensive uses. This pattern, combined with the overall vegetation loss, suggests increasing pressure on the regional ecosystem. The slight increase in water bodies from 17.25 km^2 (1.37%) to 19.36 km^2 (1.54%) may reflect the creation of new water infrastructure to support the expanding urban population, though this modest increase is insufficient to offset the broader environmental impacts of urban expansion.

Dramatic Urban Expansion

The most striking transformation is evident in the built-up area category, which expanded dramatically from 64.17 km^2 (5.11%) in 1990 to 226.68 km^2 (18.05%) in 2025 (Fig. 5.6).

This represents a nearly fourfold increase, with an additional 162.51 km^2 of land converted to urban use. This expansion rate of approximately 253% over 35 years demonstrates the rapid pace of urbanization in Siliguri, likely driven by its strategic location as a gateway to the northeastern states of India and its role as a commercial hub. The LULC analysis reveals a landscape undergoing rapid transformation, with urban expansion being the dominant driver of change. The substantial loss of vegetation cover, coupled with dramatic urban growth, presents significant challenges

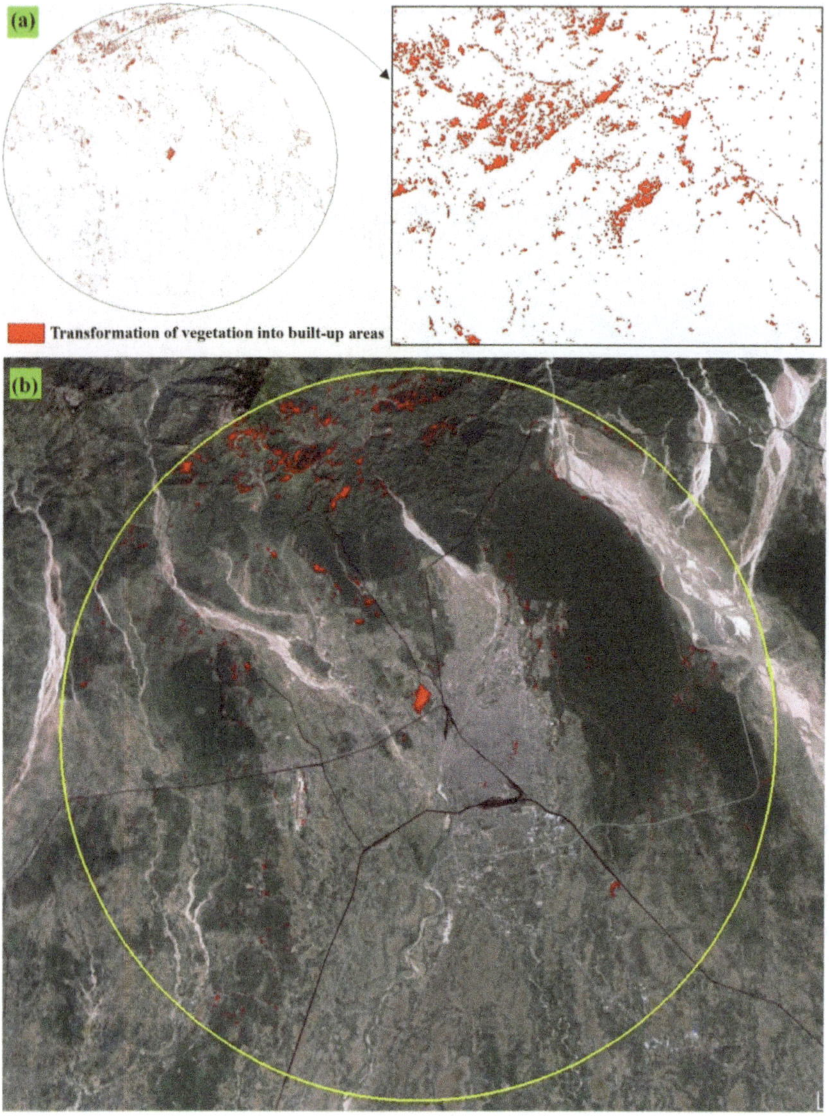

Fig. 5.6 Land transformation from vegetation to built-up area from 1990 to 2025 at Siliguri Town and surrounding areas (20 km) of West Bengal

for environmental sustainability and ecosystem service provision (Fig. 5.7). These findings underscore the urgent need for integrated land-use planning that balances urban development needs with environmental conservation objectives in the Siliguri region.

Fig. 5.7 Symmetrical diagram of restoring Siliguri Forest Corridor

5.4.3 Conclusion

The findings of Chap. 5 highlight the intensifying spatial and ecological tensions between rapid urban growth and forest ecosystems in Siliguri and its surrounding regions of West Bengal. Over the 35-year study period (1990–2025), the region has witnessed a pronounced transformation in land use and land cover, driven largely by an accelerating pace of urbanization and infrastructural development. These transformations underscore the broader socio-ecological conflicts that are becoming increasingly prevalent across urbanizing landscapes in the Global South. The most significant change has been the exponential increase in built-up areas, expanding from 64.17 km^2 to 226.68 km^2 an increase of over 250%. This dramatic urban growth, concentrated in peri-urban zones, has occurred largely at the expense of vegetated and marginal lands. The corresponding decline in vegetation cover from 490.14 km^2 to 433.85 km^2 reflects not only physical deforestation but also a degradation of critical ecosystem services, including carbon sequestration, biodiversity support, flood regulation, and groundwater recharge. The conversion of 76.42 km^2 of vegetation to agricultural land further reveals the dual pressures exerted by both urban and rural land demands.

The patterns observed in Siliguri resonate with global studies indicating that urban expansion is now among the top drivers of forest loss, particularly in biodiversity-rich and economically developing regions. This trend is particularly concerning in ecologically sensitive areas like the Terai-Duar region, where forest fragmentation has disrupted elephant corridors and increased human-wildlife conflict. These spatial transformations have not occurred in isolation they are deeply entangled with socio-economic forces, including land speculation, weak environmental governance, and inadequate planning frameworks that fail to integrate ecological priorities.

Furthermore, the evidence of ecological marginalization where forest-dependent communities and vulnerable populations are disproportionately impacted highlights a growing social justice concern. The displacement of indigenous groups and erosion of customary land tenure systems, as seen in other regions of West Bengal and India, calls for an inclusive approach to urban planning that acknowledges and protects the rights of local stakeholders. Climate change acts as a compounding stressor, exacerbating the vulnerabilities created by unplanned urban expansion. Events such as the 2017 floods in Siliguri, worsened by deforestation in watershed areas, illustrate how ecological degradation can feedback into urban vulnerability. Forest ecosystems that once served as natural buffers are now increasingly fragmented, reducing their ability to support urban resilience.

From a policy and planning perspective, these findings reinforce the urgent need for integrated landscape governance that bridges the institutional gap between urban development and forest conservation. Green infrastructure planning, ecological zoning, urban growth boundaries, and restoration of critical wildlife corridors are among the key strategies that should be prioritized. Advanced geospatial tools and satellite-based monitoring systems, as employed in this study, must continue to inform decision-making processes by providing timely, accurate data on land-cover transitions. In conclusion, the case of Siliguri exemplifies the complex and often conflict-ridden interface between urban expansion and forest conservation. As cities like Siliguri continue to grow, the challenge lies not in halting development, but in managing it wisely through strategies that integrate ecological knowledge, community participation, and sustainable land-use policies. Addressing this challenge is essential not only for safeguarding the remaining forest ecosystems of West Bengal but also for ensuring the long-term resilience and liveability of its urban landscapes.

References

Atef I, Ahmed W, Abdel-Maguid RH (2024) Future land use land cover changes in El-Fayoum governorate: a simulation study using satellite data and CA-Markov model. Stoch Env Res Risk Assess 38(2):651–664

Ali G, Mijwil MM, Adamopoulos I, Ayad J (2025) Leveraging the internet of things, remote sensing, and artificial intelligence for sustainable forest management. Babylonian J Internet Things 2025:1–65

Banerjee-Guha S (2008) Space relations of capital and significance of new economic enclaves: SEZs in India. Econ Polit Wkly 51–59

Barona CO (2015) Adopting public values and climate change adaptation strategies in urban forest management: a review and analysis of the relevant literature. J Environ Manage 164:215–221

Census of India (2011) District census handbook. Bihar Series 11 Patna, Begusarai, Smastipur and Banka

Costanza R, De Groot R, Sutton P, Van der Ploeg S, Anderson SJ, Kubiszewski I et al (2014) Changes in the global value of ecosystem services. Glob Environ Chang 26:152–158

Curtis PG, Slay CM, Harris NL, Tyukavina A, Hansen MC (2018) Classifying drivers of global forest loss. Science 361(6407):1108–1111

Desa UN (2018) 68% of the world population projected to live in urban areas by 2050, says UN. United Nations Department of Economic and Social Affairs

Du P, Liu S, Xia J, Zhao Y (2013) Information fusion techniques for change detection from multi-temporal remote sensing images. Inf Fusion 14(1):19–27

Forest Survey of India (2021) India State of Forest Report 2021. Ministry of Environment, Forest and Climate Change, Government of India. https://fsi.nic.in/isfr2021/

Grimm NB, Faeth SH, Golubiewski NE, Redman CL, Wu J, Bai X, Briggs JM (2008) Global change and the ecology of cities. Science 319(5864), 756–760

Gorelick N, Hancher M, Dixon M, Ilyushchenko S, Thau D, Moore R (2017) Google Earth Engine: Planetary-scale geospatial analysis for everyone. Remote Sens Environ 202:18–27

Hansen MC, Potapov PV, Moore R, Hancher M, Turubanova SA, Tyukavina A et al (2013) High-resolution global maps of 21st-century forest cover change. Science 342(6160):850–853

Iyer KG (2020) Tribal land, forest and identity issues: a study of Jharkhand and Odisha. Concept Publishing Company

Koko AF, Han Z, Wu Y, Abubakar GA, Bello M (2022) Spatiotemporal land use/land cover mapping and prediction based on hybrid modeling approach: a case study of Kano Metropolis, Nigeria (2020–2050). Remote Sens 14(23):6083

Laurance WF, Nascimento HE, Laurance SG, Andrade A, Ewers RM, Harms KE et al (2007) Habitat fragmentation, variable edge effects, and the landscape-divergence hypothesis. PLoS ONE 2(10):e1017

Larson AM (ed) (2010) Forests for people: community rights and forest tenure reform. Earthscan

McGarigal K, Cushman SA, Neel MC, Ene E (2002) FRAGSTATS: spatial pattern analysis program for categorical maps. Computer software program produced by the authors at the University of Massachusetts, Amherst, 3

Mukhopadhyay A, Hornby DD, Hutton CW, Lázár AN, Amoako Johnson F, Ghosh T (2018) Land cover and land use analysis in coastal Bangladesh. Ecosystem services for well-being in deltas: Integrated assessment for policy analysis. Springer International Publishing, Cham, pp 367–381

Nandy J, Dey S, Heilimo E (2019) Neoarchaean magmatism through arc and lithosphere melting: evidence from eastern Dharwar Craton. Geol J 54(5):3148–3166

Obidzinski K, Andriani R, Komarudin H, Andrianto A (2012) Environmental and social impacts of oil palm plantations and their implications for biofuel production in Indonesia. Ecol Soc 17(1)

Panja U, Mistri B (2025) Human-elephant conflict risks in the forest-dominated areas of West Bengal, India. Environ Monit Assess 197(6):659

Pelletier C, Valero S, Inglada J, Champion N, Dedieu G (2016) Assessing the robustness of Random Forests to map land cover with high resolution satellite image time series over large areas. Remote Sens Environ 187:156–168

Pradhan A, Khaling S (2024) Structure, composition, and regeneration status of trees in non-protected forests of Eastern Himalaya. Trop Ecol 65(3):443–459

Polgar G, Jaafar Z (2018) Endangered forested wetlands of Sundaland. Springer, Berlin Heidelberg, pp 89–93

Rossi F, Becker G (2019) Creating forest management units with Hot Spot Analysis (Getis-Ord Gi*) over a forest affected by mixed-severity fires. Aust for 82(4):166–175

Roy PS, Meiyappan P, Joshi PK, Kale MP, Srivastav VK, Srivasatava SK et al (2016) Decadal land use and land cover classifications across India, 1985, 1995, 2005. ORNL Distributed Active Archive Center (DAAC) dataset https://doi.org/10.3334/ORNLDAAC/1336

Shit PK, Pourghasemi HR, Adhikary PP, Bhunia GS, Sati VP (eds) (2021) Forest resources resilience and conflicts. Elsevier
Wang Z, Lin L, Zhang B, Xu H, Xue J, Fu Y et al (2023) Sustainable urban development based on an adaptive cycle model: a coupled social and ecological land use development model. Ecol Ind 154:110666
Zhang X, Han L, Han L, Zhu L (2020) How well do deep learning-based methods for land cover classification and object detection perform on high resolution remote sensing imagery? Remote Sens 12(3):417
Zhu Z, Wulder MA, Roy DP, Woodcock CE, Hansen MC, Radeloff VC et al (2019) Benefits of the free and open Landsat data policy. Remote Sens Environ 224:382–385

Chapter 6
Biodiversity and Ecosystem Services in Forest Landscapes

Abstract This chapter examines the critical relationships between biodiversity patterns and ecosystem service provision in the forest landscapes of Bankura District, West Bengal, situated at the biogeographical transition between the Gangetic Plains and the Chota Nagpur Plateau. Employing an integrated methodological framework combining remote sensing, GIS-based spatial analysis, and ecosystem service valuation through Benefit Transfer and Ecosystem Service Potentiality methods, the study quantifies spatiotemporal changes in forest ecosystems over a 34-year period (1990–2024). Results reveal catastrophic ecosystem degradation, with total ecosystem service values declining from 775.11 US$ ha^{-1} yr^{-1} to 395.24 US$ ha^{-1} yr^{-1}, representing a 49% reduction in service provision capacity. Forest fragmentation analysis using NDVI-derived vegetation indices demonstrates that edge and patch forests experienced the most severe degradation, with patch forests losing 72% of their ecosystem service values, while large core areas exhibited greater resilience with only 19% decline. The comprehensive biodiversity inventory documents 78 tree species, 55 shrubs and herbs, 60+ avifaunal species, and 45+ medicinal plants, highlighting the district's ecological significance despite mounting anthropogenic pressures from agricultural expansion, mining activities, and infrastructure development. Spatial heterogeneity analysis across 22 Community Development blocks reveals differential degradation patterns, with Sonamukhi, Patrasayer, and Ranibandh experiencing maximum losses while Hirbandh demonstrated remarkable resilience. Findings underscore that forest structural integrity fundamentally determines ecosystem service stability, emphasizing the urgent necessity for landscape-scale conservation strategies integrating biodiversity protection, habitat restoration, community-based governance, and traditional ecological knowledge systems to reverse ecological decline and secure sustainable forest ecosystem functioning for forest-dependent communities and regional environmental stability.

Keywords Biodiversity and ecosystem services · Forest degradation · Remote sensing and GIS · Landscape ecology · NDVI

P. K. Shit and M. Hasanuzzaman, *Forest Landscape Dynamics in West Bengal*, Advances in Asian Human-Environmental Research,
https://doi.org/10.1007/978-3-032-23700-2_6

6.1 Introduction

Forest landscapes represent some of Earth's most complex and biologically diverse ecosystems, harboring approximately 80% of terrestrial biodiversity while occupying roughly 31% of the global land surface (FAO 2020). These intricate ecological systems function as dynamic mosaics where biodiversity and ecosystem services are inextricably linked through multifaceted relationships that operate across multiple spatial and temporal scales (Correia and Lopes 2023). Understanding these connections has become increasingly critical as anthropogenic pressures continue to fragment and degrade forest ecosystems worldwide, threatening both biological diversity and the essential services upon which human societies depend (Morris 2010). The conceptual framework linking biodiversity to ecosystem services in forest landscapes emerged from decades of ecological research demonstrating that biological diversity serves as a fundamental driver of ecosystem functioning (Tilman et al. 2001). This relationship operates through several mechanisms: species diversity enhances resource utilization efficiency, provides functional redundancy that stabilizes ecosystem processes, and creates complex interaction networks that support resilience to environmental perturbations (Loreau et al. 2001). Within forest ecosystems, these principles manifest across multiple organizational levels, from genetic diversity within tree populations to landscape-scale heterogeneity in forest structure and composition.

Forest biodiversity encompasses multiple dimensions that collectively determine ecosystem service provision. Alpha diversity, reflecting species richness within local forest communities, influences processes such as nutrient cycling, carbon sequestration, and primary productivity through complementary resource use and facilitative interactions among species (Cardinale et al. 2012). Beta diversity, representing compositional turnover across environmental gradients, contributes to landscape-scale resilience by ensuring that ecosystem functions can be maintained even when local disturbances affect specific forest patches. Gamma diversity, encompassing regional species pools, provides the evolutionary and ecological context within which local forest communities assemble and respond to environmental change (Arellano and Halffter 2003).

The ecosystem services framework, as articulated by the Millennium Ecosystem Assessment (2005), provides a structured approach to understanding the multiple ways forest biodiversity contributes to human well-being. Provisioning services, including timber, non-timber forest products, and genetic resources, depend directly on the diversity of forest species and their functional traits. Regulating services, such as climate regulation through carbon storage, water purification, and natural pest control, emerge from complex ecological processes that are enhanced by biodiversity through complementarity effects and stability mechanisms (Naeem et al. 2009). Cultural services, encompassing recreational, spiritual, and aesthetic values, are often closely tied to the presence of charismatic species and diverse forest communities that provide rich sensory experiences. Supporting services represent the ecological foundation upon which all other forest ecosystem services depend (Melaku and Pastor

Ivars 2024). Soil formation, nutrient cycling, and primary productivity are fundamental processes that maintain forest ecosystem integrity and enable the provision of direct benefits to human societies. These processes are strongly influenced by the diversity of decomposer organisms, mycorrhizal networks, and plant functional groups that together create the biogeochemical cycles essential for forest ecosystem functioning (Bardgett and Van Der Putten 2014).

Contemporary forest landscapes face unprecedented challenges from climate change, habitat fragmentation, invasive species, and intensified resource extraction (Noulèkoun et al. 2021). These pressures create complex feedbacks between biodiversity loss and ecosystem service degradation, potentially triggering cascading effects that compromise both ecological integrity and human well-being. Understanding these dynamics requires integrating perspectives from community ecology, landscape ecology, and social-ecological systems science to develop comprehensive frameworks for forest conservation and management. This chapter explores the intricate relationships between biodiversity and ecosystem services in forest landscapes, examining how biological diversity at multiple scales influences ecosystem functioning and service provision. Through synthesis of current research and theoretical frameworks, we aim to provide insights that can inform evidence-based strategies for maintaining both biodiversity and ecosystem services in an era of rapid environmental change.

6.2 Biodiversity and Ecosystem Services in Forest Landscapes of West Bengal

West Bengal, situated in the eastern region of India, encompasses diverse forest landscapes that represent critical biodiversity hotspots and provide essential ecosystem services to over 91 million inhabitants (Das 2021). The state's forest cover spans approximately 16,902 km^2, constituting 18.98% of its geographical area, and harbors exceptional biological diversity across multiple forest types ranging from tropical moist deciduous forests in the south to temperate forests in the northern hills (Forest Survey of India 2021). These forest ecosystems serve as vital corridors connecting the Eastern Himalayas and Indo-Burma biodiversity hotspots, making West Bengal a significant biogeographical junction for species conservation and ecosystem service provision.

The state's forest landscapes exhibit remarkable ecological heterogeneity, encompassing the Sundarbans mangrove forests, the Terai-Duar foothill forests, the Darjeeling Himalayan forests, and the Chota Nagpur Plateau forests. Each ecosystem type supports distinct assemblages of flora and fauna while providing unique combinations of ecosystem services (Chatterjee et al. 2006). The Sundarbans, recognized as a UNESCO World Heritage Site, represents the world's largest mangrove forest ecosystem, supporting 334 plant species and serving as critical habitat for the endangered Royal Bengal Tiger (Panthera tigris tigris) while providing coastal protection and fisheries resources (Gopal and Chauhan 2006). The northern hill forests of Darjeeling and Kalimpong districts harbor over 2500 plant species, including numerous endemic orchids and medicinal plants, while regulating water flow for major river systems (Rahamtulla et al. 2020).

The intricate relationship between biodiversity and ecosystem services in West Bengal's forests operates through complex ecological processes that have evolved over millennia (Mallick and Chakraborty 2018). Species diversity enhances ecosystem stability and resilience, particularly crucial given the region's vulnerability to climate change and extreme weather events. The state's forests contribute significantly to carbon sequestration, storing approximately 87.5 million tons of carbon, while supporting livelihoods through timber, non-timber forest products, and ecotourism (Amusa et al. 2024). Traditional ecological knowledge systems of indigenous communities, including the Lepcha, Bhutia, and various tribal groups, have historically maintained sustainable relationships with forest biodiversity, creating mosaic landscapes that optimize both conservation and human well-being (Pattnaik 2024).

Contemporary challenges facing West Bengal's Forest biodiversity stem from multiple anthropogenic pressures including urbanization, agricultural expansion, infrastructure development, and climate change. The state has lost approximately 1200 km^2 of forest cover since 1990, with fragmentation effects cascading through ecosystem service provision (Sudhakar Reddy et al. 2016). Tea plantation expansion in the hills, shrimp farming in coastal areas, and urban sprawl around Kolkata have created significant habitat pressures, while climate change threatens species distributions and ecosystem functioning patterns (Prasad and Das 2025).

The Government of West Bengal has implemented various conservation initiatives, including the establishment of 15 wildlife sanctuaries, 6 national parks, and multiple reserved forests covering 5224 km^2 (Das 2021). However, effective conservation requires understanding the complex interactions between biodiversity patterns and ecosystem service flows across different forest types and management regimes. Recent research emphasizes the need for landscape-scale approaches that integrate biodiversity conservation with sustainable development goals (Mukhopadhyay et al. 2015). This chapter examines the multifaceted relationships between biodiversity and ecosystem services across West Bengal's diverse forest landscapes, analyzing how ecological processes, human activities, and conservation strategies interact to shape current and future sustainability outcomes.

6.3 Biodiversity and Ecosystem Services in Forest Landscapes-Calculation Tools and Techniques in GIS

The quantification and spatial analysis of biodiversity and ecosystem services in forest landscapes have been revolutionized by advances in Geographic Information Systems (GIS) technology and remote sensing applications (Nizamani et al. 2024) (Fig. 6.1). Modern conservation science increasingly relies on sophisticated computational tools to map, model, and monitor the complex relationships between forest biodiversity patterns and ecosystem service provision across multiple spatial and temporal scales (Cord et al. 2017). These technological advances have enabled researchers and forest managers to transition from qualitative assessments to quantitative, spatially explicit analyses that support evidence-based decision-making in forest conservation and management.

GIS-based calculation tools for biodiversity assessment encompass a diverse array of spatial analytical techniques designed to quantify species distribution patterns, habitat suitability, and landscape connectivity (Böhm and Popescu 2016). Species Distribution Models (SDMs) utilizing algorithms such as Maximum Entropy (MaxEnt) and Random Forest have become fundamental tools for predicting species occurrence probabilities across forest landscapes using environmental variables derived from remote sensing data (Elith et al. 2011). Landscape metrics, including patch size distribution, edge density, and connectivity indices, provide quantitative measures of habitat fragmentation and its impacts on biodiversity maintenance (McGarigal et al. 2002). Alpha, beta, and gamma diversity indices can be calculated spatially using field survey data integrated with satellite imagery to assess biodiversity patterns at multiple scales.

Ecosystem service quantification in forest landscapes has been significantly enhanced through the development of specialized GIS-based modeling platforms. The Integrated Valuation of Ecosystem Services and Tradeoffs (InVEST) toolkit provides spatially explicit models for carbon storage, water yield, habitat quality, and pollination services, enabling researchers to map ecosystem service provision

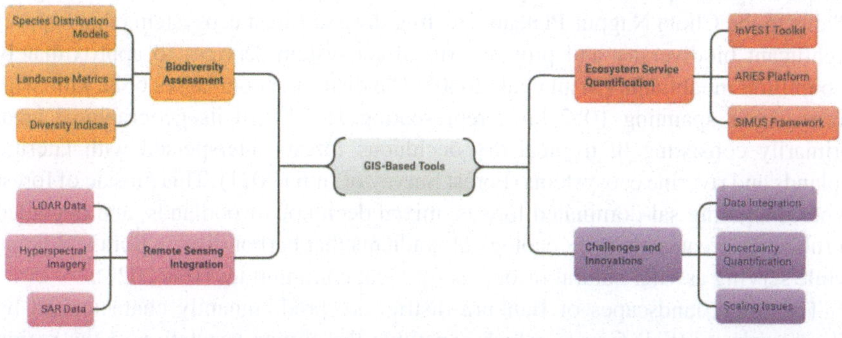

Fig. 6.1 Flow diagram of GIS-based tools for forest biodiversity and ecosystem services

across heterogeneous forest landscapes (Sharp et al. 2020). The Artificial Intelligence for Ecosystem Services (ARIES) platform employs machine learning algorithms to quantify ecosystem service flows and beneficiary mapping, while the System for Integrated Modeling of Urban Systems (SIMUS) framework addresses urban-forest interface dynamics. Remote sensing integration has become essential for large-scale biodiversity and ecosystem service assessments. Light Detection and Ranging (LiDAR) data enables precise forest structure characterization, supporting biomass estimation and habitat complexity analyses. Hyperspectral imagery facilitates species identification and functional diversity mapping, while synthetic aperture radar (SAR) data provides all-weather monitoring capabilities for forest change detection (Fassnacht et al. 2016). Machine learning algorithms, including deep learning approaches, are increasingly employed to automate species classification and ecosystem service mapping from multi-temporal satellite datasets. Contemporary challenges in GIS-based forest biodiversity and ecosystem service assessment include data integration across multiple sources, uncertainty quantification in model outputs, and scaling issues between field measurements and satellite observations. Cloud computing platforms such as Google Earth Engine have democratized access to large-scale geospatial analysis capabilities, enabling researchers to process massive datasets for continental-scale assessments (Gorelick et al. 2017). This chapter explores the comprehensive suite of GIS calculation tools and techniques available for quantifying biodiversity and ecosystem services in forest landscapes, examining their applications, limitations, and emerging innovations in computational ecology and conservation science.

6.4 A Case Study on Biodiversity and Ecosystem Services in Forest Landscapes in Bankura District of West Bengal

Bankura district, located in the southwestern region of West Bengal, represents a unique biogeographical landscape characterized by the transition from the Gangetic Plains to the Chota Nagpur Plateau, creating diverse forest ecosystems that support significant biodiversity and provide critical ecosystem services to approximately 3.6 million inhabitants (Bisui et al. 2020). The district encompasses 6882 km^2 with forest cover spanning 1052 km^2, representing 15.29% of its geographical area, primarily consisting of tropical dry deciduous forests interspersed with lateritic uplands and riverine ecosystems (Forest Survey of India 2021). This mosaic of forest types, including sal-dominated forests, mixed deciduous woodlands, and degraded scrublands, creates complex ecological gradients that harbor diverse flora and fauna while serving as vital natural resources for local communities (Das 2021).

The forest landscapes of Bankura district are predominantly characterized by *Shorea robusta* (Sal) forests, which constitute the climax vegetation of the region and support diverse plant communities including *Terminalia tomentosa*, *Buchanania lanzan*, and *Diospyros melanoxylon* (Bandyopadhyay and Lakshminarasimhan 2010). These forests provide habitat for 47 mammalian species, 195 bird species, and numerous reptilian and amphibian species, including several threatened taxa such as the Indian pangolin (*Manis crassicaudata*) and the Asian elephant (*Elephas maximus*) in the western forest corridors (Chakraborty et al. 2013). The region's biodiversity significance is further enhanced by its position as a transitional zone between the humid eastern forests and the drier western deciduous forests, creating unique ecological niches that support both eastern and western floristic elements.

Ecosystem services provided by Bankura's forest landscapes encompass multiple dimensions of human well-being and environmental regulation. Carbon sequestration in these forests contributes approximately 2.1 million tons of stored carbon, playing a crucial role in climate regulation (Bisui and Shit 2025). Water regulation services are particularly significant, as these forests influence the hydrological regimes of major rivers including the Damodar, Rupnarayan, and Silai, supporting irrigation and drinking water supplies for rural communities. Non-timber forest products, including sal seeds, mahua flowers, kendu leaves, and medicinal plants, provide livelihood opportunities for over 200,000 tribal and marginal communities, contributing approximately ₹150 crores annually to the local economy (Gupta et al. 2021).

The district faces significant conservation challenges stemming from anthropogenic pressures and climate change impacts. Deforestation rates of 0.8% annually have been documented over the past decade, primarily driven by mining activities, agricultural expansion, and infrastructure development (Alam et al. 2022). The presence of coal mining operations in the eastern districts has created habitat fragmentation, affecting wildlife corridors and ecosystem connectivity. Additionally, climate change projections indicate increasing temperature and altered precipitation patterns that may shift forest composition and species distributions (Mandal and Das Chatterjee 2023). Traditional ecological knowledge systems of indigenous communities, particularly the Santhal, Munda, and Bhumij tribes, have historically maintained sustainable forest management practices through community-based conservation initiatives. These traditional systems, combined with modern conservation approaches, offer valuable insights for developing integrated management strategies that balance biodiversity conservation with sustainable livelihood generation (Singhal et al. 2021). This case study examines the complex relationships between biodiversity patterns and ecosystem service provision in Bankura district's forest landscapes, analyzing spatial variations in forest structure, species composition, and service flows while evaluating the effectiveness of current conservation strategies and proposing evidence-based management recommendations for sustainable forest ecosystem management.

6.4.1 Materials and Methods

Data Sources and Image Acquisition

For analyzing long-term land cover dynamics and ecosystem characteristics, Landsat satellite imagery was utilized due to its proven effectiveness in time-series environmental monitoring (Vogelmann et al. 2012). Landsat Thematic Mapper (TM) and Operational Land Imager (OLI) datasets with a spatial resolution of 30 m were acquired from the United States Geological Survey (USGS) Earth Explorer portal. Imagery for the years 1990 and 2024 was selected to represent the temporal spectrum of the study. These datasets served as the foundation for Normalized Difference Vegetation Index (NDVI) analysis and further land cover classification.

Forest Cover Extraction Using NDVI

The NDVI was derived from the selected satellite imagery to identify and delineate vegetation cover. This index, widely recognized in remote sensing for vegetation monitoring, compares the reflectance of near-infrared (NIR) and red bands to evaluate vegetation health. The NDVI is calculated as

$$\mathrm{NDVI} = \frac{(\mathrm{NIR} - \mathrm{Red})}{(\mathrm{NIR} + \mathrm{Red})} \tag{6.1}$$

The resulting NDVI values range from -1 to $+1$, where values below 0 typically indicate non-vegetated surfaces (e.g., water, snow, bare soil), values between 0 and 0.5 represent sparse to moderate vegetation, and values above 0.5 suggest dense vegetation (Pettorelli et al. 2005; Xie et al. 2008). A threshold NDVI value was determined to isolate vegetated areas for subsequent fragmentation analysis.

Forest Fragmentation Analysis

Forest fragmentation was assessed using the Landscape Fragmentation Tool (LFT) developed by the Center for Land Use Education and Research (CLEAR) at the University of Connecticut (Randhawa et al. 2024). This GIS-based tool classifies forested landscapes into six hierarchical categories based on their structural integrity and spatial context: *Patch, Edge, Perforated, small core (<250 acres), medium core (250 and 500 acres), and large core areas (>500 acres)* (Vogt et al. 2007). This classification was applied to NDVI-based vegetation maps for 1990 and 2024 to detect shifts in forest fragmentation patterns caused by anthropogenic pressures or natural processes. The ecological significance of each class was interpreted, with larger core areas being considered more ecologically stable and patches being more vulnerable.

Ecosystem Service Valuation

Ecosystem service values (ESVs) were estimated using the Benefit Transfer Method, a widely adopted approach that uses pre-established coefficients to assign monetary value to ecosystem functions (Costanza et al. 1997, 2014). This method calculates ESV as

$$\mathrm{ESV}_k = \mathrm{A}_k \times \mathrm{VC}_k \tag{6.2}$$

$$\mathrm{ESV}_t = \sum (A_k \times \mathrm{VC}_k) \tag{6.3}$$

where ESV_k is the value for a specific land cover type, A_k is the area of that type, and VC_k is the value coefficient assigned to the ecosystem function. Total ecosystem value (ESV_t) is derived by summing all individual contributions. Recognizing that standardized coefficients may not always reflect local realities—especially in resource-constrained or agriculturally driven areas—this study also considered Ecosystem Service Potentiality (ESP) by incorporating expert judgement and local stakeholder insights to improve context relevance.

6.4.2 Results

Biodiversity of the Bankura District

The Bankura District of West Bengal represents a significant biodiversity repository within the eastern Indian biogeographic landscape, characterized by its unique transitional position between the Gangetic Plains and the Chota Nagpur Plateau. Spanning across diverse ecological zones, the district encompasses approximately 1463.56 km^2 of forest area, constituting 21.27% of its total geographical extent, and supporting remarkable floristic and faunistic diversity that reflects the complex interplay of tropical deciduous forest ecosystems and anthropogenic influences. The eastern portion of the district forms part of the rice plains of West Bengal, containing usual marshy weeds of Gangetic Plain, with aquatic plants and water weeds found in ponds, ditches, and still streams, while the western uplands support extensive sal-dominated forests that harbor numerous endemics and economically important species.

The floristic composition of Bankura District demonstrates exceptional taxonomic richness, with documented records indicating the presence of over 150 vascular plant species ranging from large timber trees to medicinal herbs and climbers (Table 6.1). The comprehensive flora inventory reveals 78 tree species, 55 shrub and herb species, and 17 climber species, dominated by families such as Fabaceae, Euphorbiaceae, Combretaceae, and Moraceae. Key forest formations include tropical dry deciduous forests characterized by *Shorea robusta* (sal) as the dominant canopy species, accompanied by associates like *Terminalia arjuna, T. tomentosa, Adina cordifolia*, and *Madhuca latifolia* (Table 6.2). The understory vegetation comprises economically significant species including *Phyllanthus emblica* (amlaki), *Terminalia chebula* (haritaki), and *Terminalia belerica* (bahera), collectively forming the traditional "Triphala" complex, alongside numerous other medicinal plants such as *Asparagus racemosus, Rauwolfia serpentina*, and *Tinospora cordifolia*. Research on ethnomedicinal plants reveals that Malvaceae, Asteraceae, Amaranthaceae, and Euphorbiaceae were recorded as dominant families during botanical surveys, highlighting the district's role as a repository of traditional pharmaceutical resources (Table 6.3).

The faunal diversity of Bankura District exhibits remarkable adaptations to the lateritic landscape and seasonal climatic variations, with wildlife attractions including Rock Agama, different Owls, Thrushes, Indian Pitta, Paradise Flycatcher, different Bee-Eaters, Black Eagle, Munias, with rocky sal forest and grasses being home to different small mammals. The avifaunal assemblage represents a diverse spectrum of feeding guilds and habitat preferences, from forest-dependent species like the Paradise Flycatcher (*Terpsiphone paradisi*) and Indian Pitta (*Pitta brachyura*)

Table 6.1 Biodiversity inventory of Bankura District

Taxonomic group	Number of species	Dominant families/orders	Conservation status	Economic importance
Flora				
Trees	78	Fabaceae, Combretaceae, Moraceae, Euphorbiaceae	Mixed (Common to vulnerable)	Timber, medicine, NTFP
Shrubs and herbs	55	Asteraceae, Euphorbiaceae, Amaranthaceae	Mostly common	Medicine, fodder
Climbers/ Lianas	17	Fabaceae, Asclepiadaceae, Convolvulaceae	Common to rare	Medicine, construction
Medicinal plants	45+	Multiple families	Variable	Traditional healthcare
Fauna				
Avifauna	60+	Passeriformes, Coraciiformes, Accipitriformes	Common to near threatened	Ecosystem services
Mammals (Small)	25+	Chiroptera, Rodentia, Soricomorpha	Common to vulnerable	Ecological balance
Reptiles	15+	Squamata, Testudines	Common	Pest control
Herpetofauna	20+	Anura, Squamata	Common to vulnerable	Ecological indicators

Table 6.2 Forest type classification and associated biodiversity of Bankura District

Forest type	Area coverage	Dominant species	Associated fauna	Conservation priority	Forest type
Sal forests	60%	*Shorea robusta*, *Terminalia* spp.	Paradise Flycatcher, Pitta	High	Sal forests
Mixed deciduous	25%	*Madhuca latifolia*, *Diospyros* spp.	Various songbirds, small mammals	Medium	Mixed deciduous
Degraded forests	15%	Secondary growth species	Generalist species	Restoration priority	Degraded forests

Table 6.3 Key species medicinal plant diversity of Bankura District

Scientific name	local name	Family	Parts used	Traditional uses
Phyllanthus emblica	Amla	Phyllanthaceae	Fruits, leaves	Antioxidant, vitamin C
Terminalia chebula	Haritaki	Combretaceae	Fruits	Digestive, anti-inflammatory
Asparagus racemosus	Satamuli	Asparagaceae	Roots	Reproductive health
Rauwolfia serpentina	Chandra	Apocynaceae	Roots	Hypertension, sedative
Tinospora cordifolia	Gulancha	Menispermaceae	Stem	Immunomodulator
Calotropis gigantea	Akanda	Apocynaceae	Leaves, latex	Wound healing, respiratory

to open woodland specialists such as various bee-eater species (*Merops* spp.) and the Black Eagle (*Ictinaetus malaiensis*). The herpetofauna is notably represented by the Rock Agama (*Psammophilus dorsalis*), adapted to the characteristic lateritic outcrops that define much of the district's topography. Small mammalian fauna includes various species of shrews, bats, rodents, and possibly small carnivores, though comprehensive mammalian surveys remain limited. Ethnobotanical studies identify key medicinal species including Calotropis gigantea, Hemidesmus indicus, Pergularia daemia, Telosma pallida, and Tylophora indica as frequently used species for veterinary and human healthcare, indicating the persistence of traditional ecological knowledge systems.

The ecological significance of Bankura's biodiversity extends beyond mere species inventories to encompass critical ecosystem services including carbon sequestration, soil conservation, watershed protection, and climate regulation. The district's forest ecosystems play a vital role in maintaining regional hydrological cycles, particularly important given the monsoonal climate and the presence of significant river systems including the Damodar and its tributaries. However, contemporary biodiversity faces mounting pressures from agricultural expansion, urbanization, mining activities, and climate change impacts, necessitating integrated conservation strategies that balance developmental needs with ecological sustainability. The presence of traditional agroforestry systems, sacred groves, and community-managed forest patches provides valuable models for sustainable biodiversity conservation, while ongoing efforts in afforestation and habitat restoration offer hope for maintaining the district's rich natural heritage for future generations.

Ecosystem Service

Temporal Analysis of Ecosystem Service Values in Bankura District

The comprehensive assessment of ecosystem service values (ESV) across Bankura District demonstrates a pronounced declining trend over the 34-year study period from 1990 to 2024 (Fig. 6.2). The district-wide total ESV experienced a substantial decrease from 775.11 US\$ ha^{-1} yr^{-1} in 1990 to 395.24 US\$ ha^{-1} yr^{-1} in 2024, representing an absolute loss of 379.87 US\$ ha^{-1} yr^{-1}. This decline constitutes approximately 49% reduction in ecosystem service provision capacity over the study period (Table 6.4).

The temporal analysis reveals a consistent pattern of ecosystem service degradation across the district, with no blocks showing improvement in ESV over the study period. This uniform declining trend indicates widespread environmental pressures affecting the entire district's ecological integrity and service provision capacity.

Spatial Distribution of ESV Changes Across CD Blocks

The spatial analysis of ecosystem service value changes across the 22 Community Development (CD) blocks in Bankura District reveals significant heterogeneity in the magnitude of ESV losses (Table 6.4 and Fig. 6.3). The blocks can be categorized into different degradation intensity classes based on their ESV loss patterns.

Several blocks experienced exceptionally high ESV losses during the study period. Sonamukhi recorded the maximum absolute loss of 47.33 US\$ ha^{-1} yr^{-1}, declining from 84.03 to 36.70 US\$ ha^{-1} yr^{-1}. Patrasayer followed with a loss of 30.73 US\$ ha^{-1} yr^{-1} (from 42.73 to 12.00 US\$ ha^{-1} yr^{-1}), while Ranibandh experienced a decline of 29.07 US\$ ha^{-1} yr^{-1} (from 93.59 to 64.52 US\$ ha^{-1} yr^{-1}). Indus and Barjora also showed substantial losses of 27.76 and 27.55 US\$ ha^{-1} yr^{-1}, respectively.

A significant number of blocks experienced moderate levels of ESV degradation. Kotulpur lost 25.69 US\$ ha^{-1} yr^{-1}, while Bishnupur and Onda recorded losses of 23.38 and 23.17 US\$ ha^{-1} yr^{-1}, respectively. Joypur experienced a decline of 22.08 US\$ ha^{-1} yr^{-1}, and Gangajalghati showed a loss of 17.04 US\$ ha^{-1} yr^{-1}.

Several blocks demonstrated relatively lower rates of ESV decline compared to the district average. Hirbandh recorded the minimum loss of 1.17 US\$ ha^{-1} yr^{-1} (from 11.26 to 10.09 US\$ ha^{-1} yr^{-1}), representing the most resilient block in terms

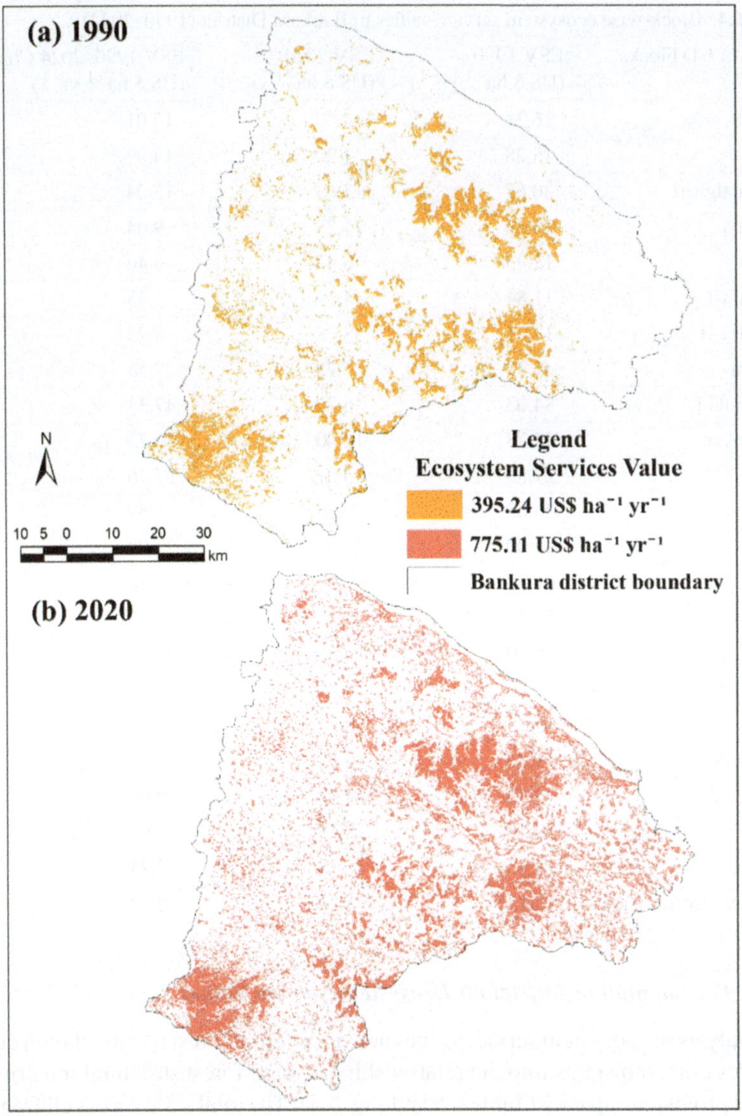

Fig. 6.2 Temporal changes in ecosystem service value (ESV) for Bankura District: A 1990–2024 analysis using benefit transfer and ecosystem service potentiality

of ecosystem service maintenance. Indpur and Khatra showed modest losses of 6.46 and 6.58 US$ ha^{-1} yr^{-1}, respectively. Bankura-I (7.25 US$ ha^{-1} yr^{-1}), Sarenga (7.75 US$ ha^{-1} yr^{-1}), and Taldangra (8.32 US$ ha^{-1} yr^{-1}) also fell into this category.

Table 6.4 Block-wise ecosystem service values in Bankura District (1990–2024)

Name of CD blocks	ESV 1990 (US \$ ha^{-1} yr^{-1})	ESV 2024 (US \$ ha^{-1} yr^{-1})	ESV 1990–2024 Change (US \$ ha^{-1} yr^{-1})
Saltora	16.38	4.37	12.01
Mejia	15.28	1.20	14.09
Gangajalghati	30.67	13.63	17.04
Chhatna	20.30	11.26	9.04
Indpur	12.93	6.47	6.46
Bankura-I	11.84	4.59	7.25
Bankura-II	17.71	8.98	8.73
Barjora	61.33	33.78	27.55
Sonamukhi	84.03	36.70	47.33
Patrasayer	42.73	12.00	30.73
Indus	27.83	0.07	27.76
Kotulpur	26.05	0.36	25.69
Joypur	45.00	22.92	22.08
Bishnupur	61.32	37.93	23.38
Onda	51.37	28.20	23.17
Taldangra	38.20	29.88	8.32
Simlapal	37.60	24.94	12.66
Khatra	15.29	8.70	6.58
Hirbandh	11.26	10.09	1.17
Ranibandh	93.59	64.52	29.07
Raipur	30.46	18.45	12.01
Sarenga	23.93	16.18	7.75
Bankura District Total	775.1131	395.242	379.87

Forest Fragmentation Impact on Ecosystem Service Values

The analysis of ecosystem service values across different forest fragmentation classes provides critical insights into the relationship between forest structural integrity and service provision capacity (Table 6.5 and Fig. 6.4). The total ESV across all fragmentation classes declined from 774.25 US\$ ha^{-1} yr^{-1} in 1990 to 394.56 US\$ ha^{-1} yr^{-1} in 2024, showing an aggregate loss of 379.69 US\$ ha^{-1} yr^{-1}. Edge forests, representing the transition zones between forested and non-forested areas, exhibited the highest absolute ESV loss among all fragmentation classes. The ESV in edge forests declined from 271.99 US\$ ha^{-1} yr^{-1} in 1990 to 114.48 US\$ ha^{-1} yr^{-1} in 2024, representing a loss of 157.51 US\$ ha^{-1} yr^{-1}. This constitutes approximately 58% of the total ESV decline across all fragmentation classes. Patch forests, characterized by isolated forest fragments, experienced the second-highest absolute ESV loss. The values decreased from 105.36 US\$ ha^{-1} yr^{-1} in 1990 to 29.05 US\$ ha^{-1} yr^{-1} in 2024, showing a decline

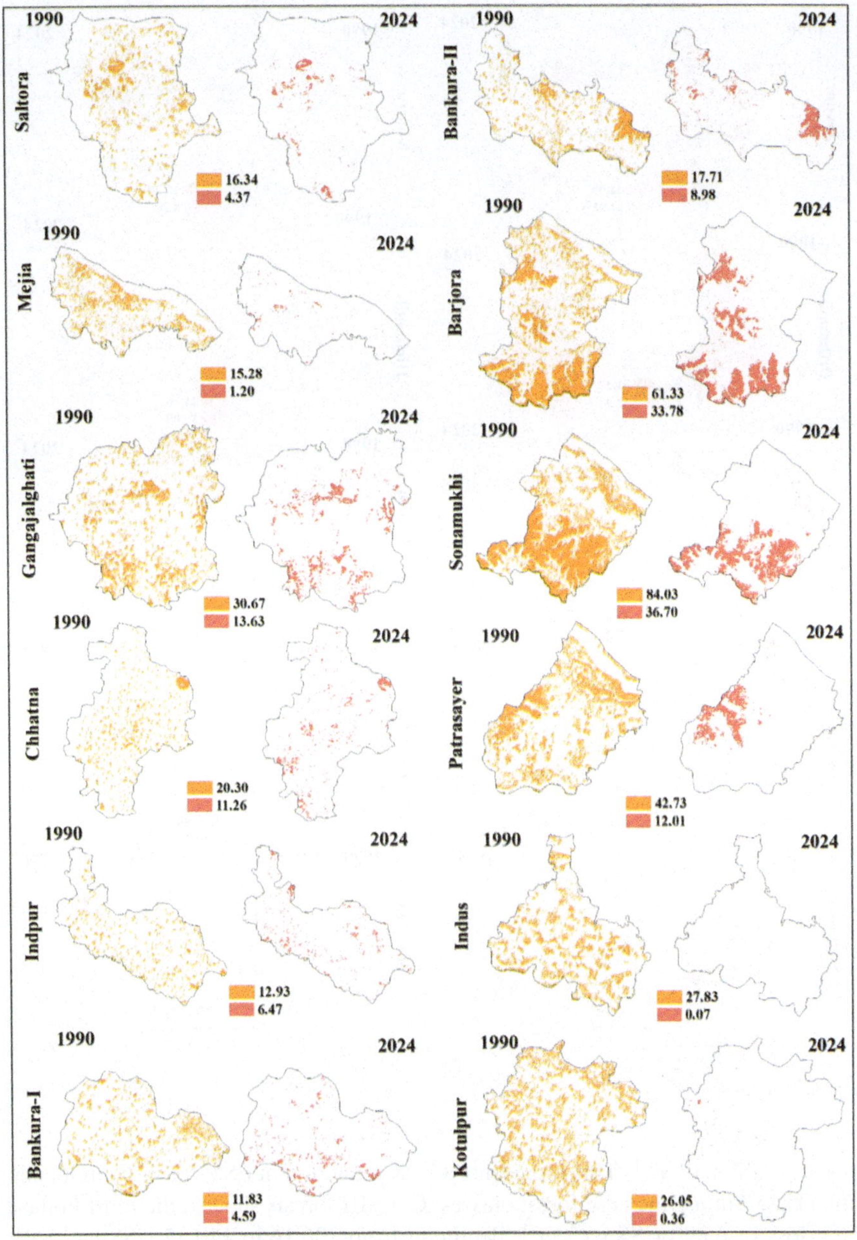

Fig. 6.3 CD Block-wise spatiotemporal dynamics of ecosystem service values (1990–2024): a benefit transfer and ecosystem service potentiality analysis for Bankura District, West Bengal

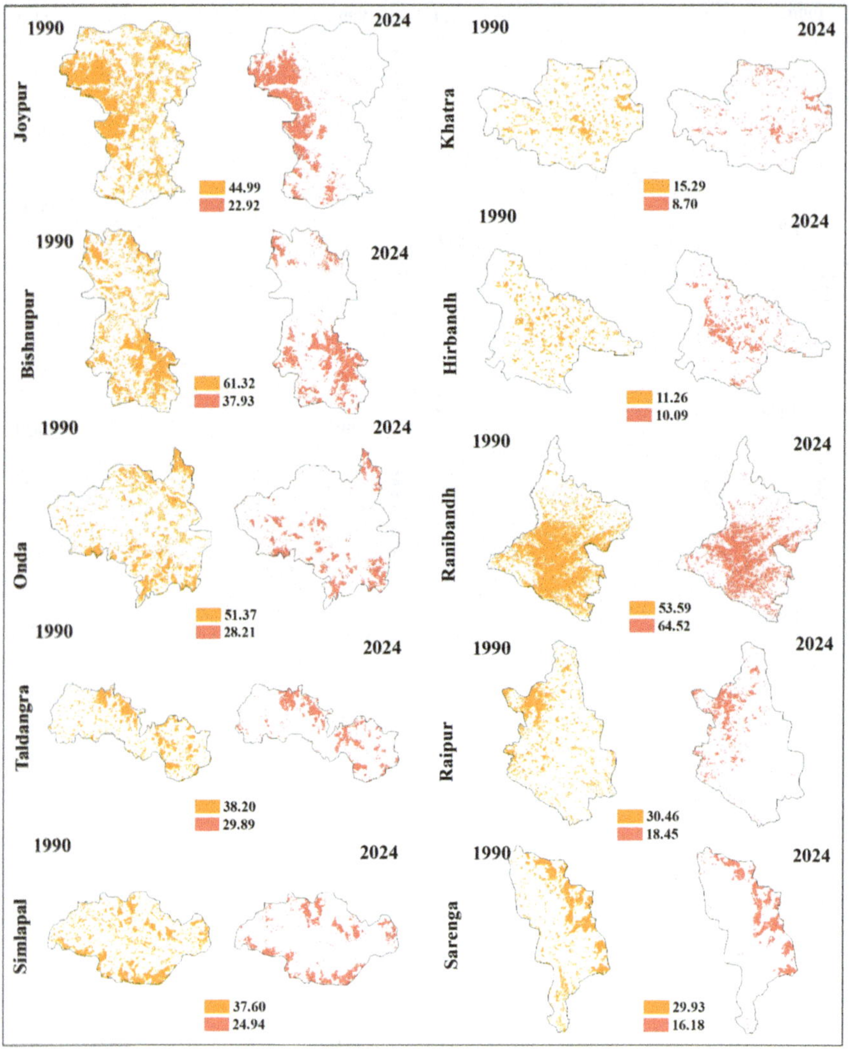

Fig. 6.3 (continued)

of 76.31 US\$ ha^{-1} yr^{-1}. This represents a 72% reduction in ESV for this fragmentation class. Among the core forest classes, Core-III forests showed the third-highest ESV loss of 65.90 US\$ ha^{-1} yr^{-1}, declining from 206.16 to 140.25 US\$ ha^{-1} yr^{-1}. Core-I forests experienced a moderate loss of 42.51 US\$ ha^{-1} yr^{-1} (from 116.51 to 74.01 US\$ ha^{-1} yr^{-1}). Core-II forests demonstrated the highest resilience among all fragmentation classes, with the smallest absolute loss of 5.36 US\$ ha^{-1} yr^{-1}, declining from 27.92 to 22.56 US\$ ha^{-1} yr^{-1}. Perforated forests, characterized by forests with internal gaps or clearings, showed a moderate ESV decline of 32.10 US$ ha^{-1} yr^{-1}, decreasing from 46.31 US$ ha^{-1} yr^{-1} in 1990 to 14.21 US$ ha^{-1} yr^{-1} in 2024.

Comparative Analysis of ESV Degradation Patterns

The comparative analysis reveals distinct patterns in ecosystem service value degradation across spatial and structural dimensions. At the block level, the coefficient of variation in ESV losses indicates high spatial heterogeneity, with losses ranging from 1.17 US$ ha^{-1} yr^{-1} to 47.33 US$ ha^{-1} yr^{-1}. This represents a 40-fold difference between the most and least affected blocks.

The forest fragmentation analysis demonstrates that structural integrity significantly influences ecosystem service stability. Core-II forests showed the highest resilience with only 19% ESV loss, while patch forests experienced the highest proportional decline of 72%. Edge forests, despite having higher absolute ESV values initially, lost 58% of their service provision capacity.

Synthesis of ESV Change Patterns

The results indicate that ecosystem service degradation in Bankura District is characterized by both spatial and structural vulnerability patterns. The data reveals that no area within the district remained immune to ESV decline, suggesting district-wide environmental pressures. However, the magnitude of change varies significantly, indicating differential resilience and vulnerability across the landscape.

The forest fragmentation analysis confirms that structural connectivity and protection status influence ecosystem service stability, with core protected areas showing greater resilience compared to fragmented and edge habitats. The consistent decline across all categories underscores the pervasive nature of environmental pressures affecting the region's ecological integrity over the past three decades.

Table 6.5 Ecosystem service values by forest fragmentation classes in Bankura District (1990–2024)

Forest fragmentation class names	ESV 1990 (US $ ha^{-1} yr^{-1})	ESV 2024 (US $ ha^{-1} yr^{-1})	ESV 1990–2024 change (US $ ha^{-1} yr^{-1})
Patch	105.36	29.05	76.31
Edge	271.99	114.48	157.51
Perforated	46.31	14.21	32.10
Core-I	116.51	74.01	42.51
Core-II	27.92	22.56	5.36
Core-III	206.16	140.25	65.90
Total	774.25	394.56	379.69

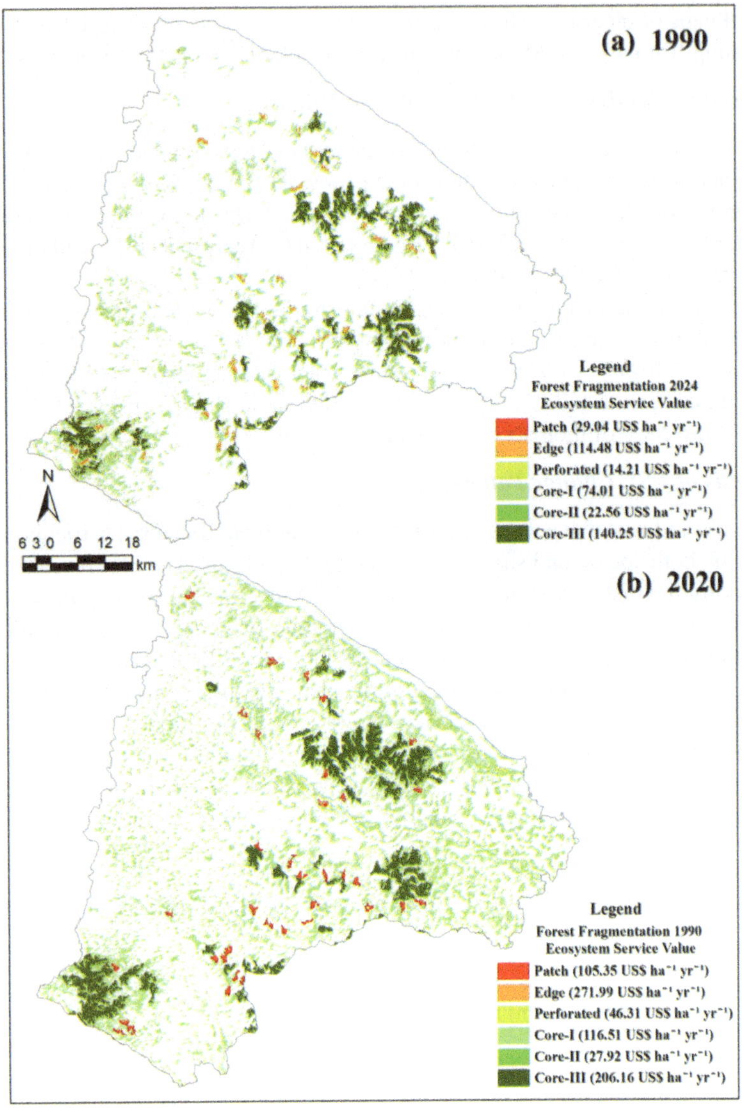

Fig. 6.4 Forest fragmentation and ecosystem service degradation in Bankura District: a 30-year spatial analysis (1990–2024)

6.4.3 *Discussion*

The findings of this study illuminate the intricate and multifaceted relationships between biodiversity patterns and ecosystem service values (ESVs) in the forested landscapes of Bankura District, West Bengal. The consistent and substantial decline in ecosystem services between 1990 and 2024 approximately 49% reduction in overall ESV reflects a growing ecological crisis driven by forest fragmentation, land-use change, and anthropogenic pressures.

Biodiversity Decline and Ecological Implications

Bankura's biodiversity, shaped by its position at the interface of the Gangetic Plains and the Chota Nagpur Plateau, is marked by high species richness and endemism, particularly within its tropical dry deciduous forests. However, the ongoing loss of biodiversity evident in reduced forest cover, degradation of sal-dominated ecosystems, and declining presence of key floral and faunal species signals a breakdown in the ecological equilibrium that supports provisioning, regulating, supporting, and cultural ecosystem services (Cardinale et al. 2012; FAO 2020).

The rich assemblage of medicinal plants, small mammals, birds, and herpetofauna provides essential ecological functions including pollination, seed dispersal, pest regulation, and traditional healthcare support. As these species diminish in abundance or distribution, the functionality and resilience of forest ecosystems are compromised. The loss of keystone species and the simplification of forest community structures may lead to trophic imbalances and a decline in ecosystem multifunctionality (Loreau et al. 2001; Naeem et al. 2009).

Spatial Patterns of Ecosystem Service Degradation

The spatial analysis reveals a sharp gradient in ESV loss across the district's CD blocks, with areas such as Sonamukhi, Patrasayer, and Ranibandh experiencing the highest reductions. These losses likely result from a combination of deforestation, agricultural encroachment, infrastructure expansion, and unsustainable resource extraction. Conversely, blocks such as Hirbandh and Indpur, which demonstrated relatively modest declines, may reflect either lower exposure to land-use pressures or more effective local conservation strategies.

The heterogeneity in degradation levels indicates that while environmental stressors are pervasive, the intensity and impact vary based on local ecological conditions, governance frameworks, and land management practices. Such variability highlights the need for geographically nuanced and site-specific conservation interventions.

Forest Fragmentation and Ecosystem Stability

The forest fragmentation assessment provides compelling evidence of the structural vulnerabilities influencing ESV loss. Edge and patch forests—subject to intense human interference and ecological edge effects exhibited the most dramatic declines in ecosystem service provision (Fig. 6.5). Patch forests, in particular, lost over 72% of their ESVs, indicating extreme ecological fragility and diminished service output.

In contrast, large core areas retained greater functional capacity, demonstrating the buffering role of interior forests against degradation.

Fig. 6.5 Field evidence of forest resource use and landscape transformation in the study area. **a** A local woman carrying collected fuelwood from a recently burned forest patch, indicating dependence on forest biomass for household energy. **b** A villager transporting bundled firewood on a bicycle from the forest, reflecting small-scale extraction and informal forest product movement. **c** Community members engaged in collective forest-based activities within a Sal-dominated woodland, showing continued human–forest interaction. **d** Post-harvest agricultural field adjacent to forest cover, illustrating the forest–agriculture interface and land-use transition. **e** A forest-edge household member with grazing livestock inside the woodland, highlighting the role of forests in supporting rural livelihoods. **f** Freshly cut and stacked timber within the forest, evidencing ongoing wood harvesting and localized forest disturbance

These results affirm the theoretical expectation that fragmentation disrupts ecosystem connectivity, alters microclimatic conditions, and reduces habitat quality (Vogt et al. 2007; Jaeger (2000). The reduced ESVs in perforated and edge forests can be attributed to increased exposure to invasive species, fire risks, soil erosion, and biodiversity loss factors that collectively diminish ecological productivity and stability.

The integration of NDVI-based vegetation analysis, GIS-based fragmentation metrics, and the Benefit Transfer and Ecosystem Service Potentiality (ESP) valuation frameworks provides a robust methodological foundation for quantifying landscape-scale ecological change. These tools enabled spatially explicit assessment of both biophysical and socio-economic dimensions of ecosystem services (Costanza et al. 1997; Sharp et al. 2020).

However, limitations persist. The Benefit Transfer method relies on generalized coefficients that may not capture local socio-ecological contexts, potentially underestimating or overestimating actual service values. Moreover, the lack of high-resolution field biodiversity data limits the ability to validate remote sensing-based assessments. Incorporating participatory mapping, long-term biodiversity monitoring, and local knowledge systems would strengthen the relevance and accuracy of future assessments.

The pervasive decline in ESVs across all forest types and administrative units signals an urgent need for integrated landscape-level conservation planning. Strategies must prioritize:

- Protection and restoration of core forest areas;
- Buffer zone management to mitigate edge degradation;
- Promotion of community-based forest governance rooted in indigenous knowledge;
- Mainstreaming biodiversity and ecosystem service valuation into regional planning and development agendas.

Additionally, agroforestry, ecological corridors, and restoration of degraded forests should be scaled up to enhance landscape connectivity and functional resilience. Conservation efforts must align with Sustainable Development Goals (SDGs), particularly those focused on climate action, biodiversity (SDG 15), and sustainable communities (SDG 11).

This study reinforces the critical role of biodiversity and forest structural integrity in sustaining ecosystem services in rapidly changing landscapes. The spatially explicit insights from Bankura District highlight both alarming trends and potential pathways for sustainable forest ecosystem management. A synthesis of ecological theory, GIS-based analytics, and local socio-ecological dynamics offers a powerful framework to inform evidence-based policy interventions aimed at reversing degradation and securing ecosystem services for future generations.

6.4.4 Conclusion

This study offers a comprehensive understanding of the complex interlinkages between biodiversity and ecosystem service values (ESVs) in the forested landscapes of Bankura District, West Bengal. Through the integration of remote sensing, GIS-based forest fragmentation analysis, and ecosystem service valuation using both Benefit Transfer and Ecosystem Service Potentiality (ESP) methods, the study reveals significant spatiotemporal degradation in ecosystem services over the 34-year period from 1990 to 2024. The sharp decline nearly 49% in overall ESV highlights the intensifying ecological strain on forest ecosystems due to anthropogenic activities, land-use conversion, and unsustainable resource extraction. Fragmentation of forest structures, especially the expansion of edge and patch classes, emerged as a critical driver of ecological vulnerability. While large core forest areas retained relatively higher functional capacity, the ecological integrity of fragmented and peripheral forest types diminished significantly. This reflects global trends emphasizing that intact and connected habitats are key to sustaining biodiversity and the multiple services ecosystems provide (Loreau et al. 2001; Vogt et al. 2007).

The biodiversity inventory of Bankura reinforces its ecological richness, with substantial representation across floral and faunal taxa, including economically and medicinally significant plant species, endemic wildlife, and herpetofauna. However, this biotic wealth is increasingly threatened by forest degradation, habitat fragmentation, and changing climatic patterns. These environmental changes have cascading effects on local livelihoods, especially for tribal and marginalized communities that rely heavily on forest-based resources. Importantly, the findings advocate for landscape-scale, site-specific conservation planning. Protecting core forest zones, restoring degraded ecosystems, and enhancing buffer and transition areas are essential to halting further ESV loss. Moreover, incorporating community-based forest governance rooted in traditional ecological knowledge can enhance resilience and stewardship at the grassroots level. The methodological framework presented here linking geospatial analysis with ecological valuation demonstrates the utility of interdisciplinary approaches in assessing and managing forest ecosystem sustainability. As India progresses toward achieving Sustainable Development Goals (SDGs), particularly SDG 13 (Climate Action) and SDG 15 (Life on Land), such integrated assessments are crucial for aligning conservation priorities with developmental planning. In conclusion, safeguarding biodiversity and ecosystem services in Bankura and similar ecologically sensitive regions demands urgent, informed, and collaborative action. The path forward lies in bridging science, policy, and local participation to build ecologically resilient and socially inclusive forest landscapes.

References

Alam A, Paul RK, Satpati L (2022) Application of remotely sensed data for estimation of indices to assess spatiotemporal aspects of droughts in Bankura District of West Bengal, India. In: Climate, environment and disaster in developing countries. Springer Nature Singapore, Singapore, pp 425–437

Amusa TO, Avana-Tientcheu ML, Awazi NP, Chirwa PW (2024) The role of non-timber forest products for sustainable livelihoods in African multifunctional landscapes. In: Trees in a sub-saharan multi-functional landscape: research, management, and policy. Springer Nature Switzerland, Cham, pp 153–178

Arellano L, Halffter G (2003) Gamma diversity: derived from and a determinant of alpha diversity and beta diversity. An analysis of three tropical landscapes. Acta Zoológica Mexicana 90:27–76

Bandyopadhyay S, Lakshminarasimhan P (2010) Bibliography and abstracts of papers on flora of West Bengal: II. Botanical Survey of India.

Bardgett RD, Van Der Putten WH (2014) Belowground biodiversity and ecosystem functioning. Nature 515(7528):505–511

Bisui S, Shit PK (2025) Effects of Sal forest replacement on carbon sequestration and soil nutrient dynamics: the study from eastern India. In: Proceedings of the Indian National Science Academy, pp 1–16

Bisui S, Roy S, Sengupta D, Bhunia GS, Shit PK (2020) Conversion of land use land cover and its impact on ecosystem services in a tropical forest. Spatial modeling in forest resources management: rural livelihood and sustainable development. Springer International Publishing, Cham, pp 589–606

Böhm, M., & Popescu, V. D. (2016). Landscape ecology, biogeography, and GIS methods. *Reptile Ecology and Conservation: A Handbook of Techniques, 298.*

Cardinale BJ, Duffy JE, Gonzalez A, Hooper DU, Perrings C, Venail P, Naeem S et al (2012) Biodiversity loss and its impact on humanity. Nature 486(7401):59–67

Chakraborty A, Joshi PK, Ghosh A, Areendran G (2013) Assessing biome boundary shifts under climate change scenarios in India. Ecol Ind 34:536–547

Chatterjee S, Saikia A, Dutta P, Ghosh D, Pangging G, Goswami AK (2006) Biodiversity significance of North east India. WWF-India, New Delhi 1–71

Cord AF, Brauman KA, Chaplin-Kramer R, Huth A, Ziv G, Seppelt R (2017) Priorities to advance monitoring of ecosystem services using earth observation. Trends Ecol Evol 32(6):416–428

Correia AM, Lopes LF (2023) Revisiting biodiversity and ecosystem functioning through the lens of complex adaptive systems. Diversity 15(8):895

Costanza R, d'Arge R, De Groot R, Farber S, Grasso M, Hannon B, Van Den Belt M (1997) The value of the world's ecosystem services and natural capital. Nature 387(6630):253–260

Costanza R, De Groot R, Sutton P, Van der Ploeg S, Anderson SJ, Kubiszewski I, Turner RK et al (2014) Changes in the global value of ecosystem services. Glob Environ Chang 26:152–158

Das GK (2021) Forests and forestry of West Bengal: survey and analysis. Springer Nature.

Elith J, Phillips SJ, Hastie T, Dudík M, Chee YE, Yates CJ (2011) A statistical explanation of MaxEnt for ecologists. Divers Distrib 17(1):43–57

FAO (2020) Food and agriculture organization. Global forest resources assessment 2020—Key findings. Rome. https://doi.org/10.4060/ca8753en. Retrieved 3 Nov 2022 from https://www.fao.org/forest-resources-assessment/2020/en/

Fassnacht FE, Latifi H, Stereńczak K, Modzelewska A, Lefsky M, Waser LT, Ghosh A et al (2016) Review of studies on tree species classification from remotely sensed data. Remote Sens Environ 186:64–87

Forest Survey of India (2021) India State of Forest Report 2021. Ministry of Environment, Forest and Climate Change, Government of India. https://fsi.nic.in/isfr2021/

Gopal B, Chauhan M (2006) Biodiversity and its conservation in the Sundarban mangrove ecosystem. Aquat Sci 68(3):338–354

Gorelick N, Hancher M, Dixon M, Ilyushchenko S, Thau D, Moore R (2017) Google earth engine: planetary-scale geospatial analysis for everyone. Remote Sens Environ 202:18–27

Gupta B, Chaudhuri P, Mishra TK (2021) Extraction of timber and nontimber forest produce of Southwest Bengal: ecology versus economy. In: Forest resources resilience and conflicts. Elsevier, pp 247–255

Jaeger JA (2000) Landscape division, splitting index, and effective mesh size: new measures of landscape fragmentation. Landscape Ecol 15(2):115–130

Loreau M, Naeem S, Inchausti P, Bengtsson J, Grime JP, Hector A, Wardle DA (2001) Biodiversity and ecosystem functioning: current knowledge and future challenges. Science 294(5543):804–808

Mallick PH, Chakraborty SK (2018) Forest, wetland and biodiversity: revealing multi-faceted ecological services from ecorestoration of a degraded tropical landscape. Ecohydrol Hydrobiol 18(3):278–296

Mandal M, Das Chatterjee N (2023) Habitat quality assessment through landscape pattern, plant species composition and landscape connectivity: case study from bankura district. Geo-Spatial analysis of forest landscape for wildlife management. Springer International Publishing, Cham, pp 57–84

McGarigal K, Cushman SA, Neel MC, Ene E (2002) FRAGSTATS: spatial pattern analysis program for categorical maps. Computer software program produced by the authors at the University of Massachusetts, Amherst, 3

Melaku A, Pastor Ivars J (2024) Urban sacred forests support human well-being through cultural ecosystem services. J Cult Heritage Manage Sustain Dev

Millennium ecosystem assessment MEA (2005) Ecosystems and human well-being, vol 5, p 563. Island press, Washington, DC

Morris RJ (2010) Anthropogenic impacts on tropical forest biodiversity: a network structure and ecosystem functioning perspective. Philosophical Trans Roy Soc B: Biol Sci 365(1558):3709–3718

Mukhopadhyay A, Mondal P, Barik J, Chowdhury SM, Ghosh T, Hazra S (2015) Changes in mangrove species assemblages and future prediction of the Bangladesh Sundarbans using Markov chain model and cellular automata. Environ Sci Process Impacts 17(6):1111–1117

Naeem S, Bunker DE, Hector A, Loreau M, Perrings C (eds) (2009) Biodiversity, ecosystem functioning, and human wellbeing: an ecological and economic perspective. OUP Oxford

Nizamani MM, Zhang Q, Muhae-Ud-Din G, Awais M, Qayyum M, Farhan M, Wang Y (2024) Application of GIS and remote-sensing technology in ecosystem services and biodiversity conservation. In: Deep learning for multimedia processing applications. CRC Press, pp 284–321

Noulèkoun F, Mensah S, Birhane E, Son Y, Khamzina A (2021) Forest landscape restoration under global environmental change: challenges and a future roadmap. Forests 12(3):276

Pattnaik BK (2024) Environmental and ecological bases of Indian traditions: search for an indigenous vision. In: Environmental and ecological sustainability through indigenous traditions: perspectives from the global south. Springer Nature Singapore, Singapore, pp 57–94

Pettorelli N, Vik JO, Mysterud A, Gaillard JM, Tucker CJ, Stenseth NC (2005) Using the satellite-derived NDVI to assess ecological responses to environmental change. Trends Ecol Evol 20(9):503–510

Prasad P, Das LK (eds) (2025) Politics of the anthropocene and climate crisis in India: seeking socio-ecological transformations. Taylor & Francis

Rahamtulla M, Roy AK, Khasim SM (2020) Orchid diversity in Darjeeling Himalaya, India: present status and conservation. In: Orchid biology: recent trends & challenges. Springer Singapore, Singapore, pp 155–188

Randhawa K, Bhattacharya P, Shimrah T (2024) Spatio-temporal analysis of urban forest fragmentation in Delhi, India, using Landscape Fragmentation Tool v2. 0. Environ Monit Assess 196(12):1273

Sharp R, Douglass J, Wolny S, Arkema K, Bernhardt J, Bierbower W, Wyatt K (2020) InVEST 3.8. 7. user's guide. The Natural Capital Project

Singhal V, Ghosh J, Bhat SS (2021) Role of religious beliefs of tribal communities from Jharkhand (India) in biodiversity conservation. J Environ Planning Manage 64(13):2277–2299

Sudhakar Reddy C, Jha CS, Dadhwal VK, Hari Krishna P, Vazeed Pasha S, Satish KV, Diwakar PG et al (2016) Quantification and monitoring of deforestation in India over eight decades (1930–2013). Biodivers Conserv 25(1):93–116

Tilman D, Reich PB, Knops J, Wedin D, Mielke T, Lehman C (2001) Diversity and productivity in a long-term grassland experiment. Science 294(5543):843–845

Vogelmann JE, Xian G, Homer C, Tolk B (2012) Monitoring gradual ecosystem change using Landsat time series analyses: case studies in selected forest and rangeland ecosystems. Remote Sens Environ 122:92–105

Vogt P, Riitters KH, Estreguil C, Kozak J, Wade TG, Wickham JD (2007) Mapping spatial patterns with morphological image processing. Landscape Ecol 22(2):171–177

Xie Y, Sha Z, Yu M (2008) Remote sensing imagery in vegetation mapping: a review. Journal of Plant Ecology 1(1):9–23

Chapter 7
Livelihood Dependencies and Forest Resource Utilization Patterns

Abstract This chapter investigates the intricate relationships between forest resource utilization patterns and livelihood dependencies in Binpur-II Community Development block, Jhargram district, West Bengal, employing a mixed-methods approach integrating household surveys, geospatial analysis, and participatory methodologies. Drawing upon primary data from 400 households across 40 villages classified by forest proximity and market accessibility, the research reveals that Non-Timber Forest Products (NTFPs) constitute 47% of total household income, establishing forests as critical economic lifelines for marginalized communities. The spatial analysis demonstrates striking geographic inequalities, with High Forest-High Road Access villages generating US $161.67 annually from forest resources compared to US $57.68 in Low Forest-Low Access settlements a threefold differential underscoring how geographical positioning fundamentally structures livelihood opportunities. Paradoxically, forest-proximate interim villages exhibit lower living standards despite maximum resource availability, deriving 60–70% of income from forests while outer villages achieve higher welfare through diversified agricultural production. The study identifies pronounced single-commodity dependencies, particularly Sal leaves (47% of NTFP income) and Mahua flowers (98.6% of interim village forest earnings), exposing households to seasonal volatility and market disruptions. Gender dynamics, educational disparities, and caste stratification compound these vulnerabilities, creating intersectional marginalization patterns. The research employs Geographic Information Systems for spatial buffering, accessibility modeling, and land-use change detection, demonstrating how geospatial technologies quantify forest-livelihood relationships. Policy implications emphasize spatially differentiated interventions integrating sustainable harvesting protocols, value-chain enhancement, market linkage development, and livelihood diversification strategies to balance conservation objectives with community welfare across West Bengal's forest-dependent landscapes.

Keywords Livelihood dependencies · Forest resource · Non-timber forest products · Geographic information systems · Jhargram

P. K. Shit and M. Hasanuzzaman, *Forest Landscape Dynamics in West Bengal*,
Advances in Asian Human-Environmental Research,
https://doi.org/10.1007/978-3-032-23700-2_7

7.1 Introduction

The intricate relationship between human communities and forest ecosystems represents one of the most complex and vital interactions in contemporary environmental and social geography (Moran and Ostrom 2005). Across the globe, approximately 1.6 billion people depend directly on forests for their livelihoods, with rural and indigenous communities demonstrating particularly profound dependencies on forest resources for subsistence, income generation, and cultural practices (Bank 2016). This dependency extends far beyond simple extraction activities, encompassing sophisticated systems of resource management, traditional ecological knowledge, and adaptive strategies that have evolved over generations (Berkes et al. 2000). Forest resource utilization patterns exhibit remarkable diversity across different geographical contexts, reflecting the complex interplay between ecological conditions, socio-economic circumstances, cultural traditions, and institutional frameworks (Pavri and Deshmukh 2003). From the agroforestry systems of sub-Saharan Africa to the non-timber forest product (NTFP) collection networks of Southeast Asian communities, these patterns reveal how human societies have developed nuanced strategies for deriving sustenance and economic value from forest ecosystems while maintaining their productive capacity (Angelsen et al. 2014).

The contemporary understanding of livelihood dependencies on forest resources has been significantly enhanced by the sustainable livelihood's framework, which emphasizes the multi-dimensional nature of poverty and the complex asset portfolios that rural households employ to secure their well-being (Ellis 2000). This framework recognizes that forest resources often serve as safety nets during times of economic stress, providing fallback options when other livelihood strategies fail, while simultaneously functioning as pathways out of poverty through income diversification and value-added activities (Vedeld et al. 2007). Recent research has revealed that forest-dependent communities typically engage with multiple forest products and services, creating diversified resource portfolios that enhance resilience against environmental and economic shocks (Awazi 2025). These portfolios commonly include timber and fuelwood, medicinal plants, food products such as fruits and nuts, materials for construction and handicrafts, and ecosystem services including water regulation and soil conservation (Shackleton et al. 2007). The relative importance of different forest products varies considerably based on factors such as forest type, accessibility, market integration, and the specific needs and capabilities of different household members.

Gender dimensions play a crucial role in shaping forest resource utilization patterns, with women and men often specializing in different types of forest activities and holding distinct knowledge systems regarding forest management. Women frequently dominate the collection of NTFPs for household consumption and local markets, while men may focus more on timber extraction and commercial hunting activities (Colfer 2013). These gendered patterns of resource use reflect broader social structures and power relations within communities, influencing both the sustainability of forest management practices and the distribution of benefits derived from forest resources (Agarwal 2015). The spatial and temporal dimensions of forest resource utilization present additional layers of complexity. Communities may utilize resources from multiple forest areas at different distances from their settlements, creating complex patterns of mobility and territorial use. Seasonal variations in resource availability and accessibility further shape utilization patterns, with communities often developing sophisticated calendars of forest-based activities that correspond to ecological cycles and agricultural schedules (Paumgarten and Shackleton 2011).

Market integration has emerged as a critical factor influencing contemporary forest resource utilization patterns. While subsistence use remains important for many communities, increasing connectivity to regional and global markets has created new opportunities for commercializing forest products, potentially transforming local resource management systems (Janse and Ottitsch 2005). However, market integration also introduces new risks and dependencies, as communities become vulnerable to price fluctuations and changing consumer preferences in distant markets (Belcher et al. 2005). The institutional context surrounding forest resource access and management significantly shapes utilization patterns and livelihood outcomes. Formal tenure systems, customary rights, government regulations, and community-based management institutions all influence how communities can access and use forest resources. The recognition and strengthening of community forest management systems have gained increasing attention as potentially effective approaches for balancing livelihood needs with conservation objectives (Agrawal and Angelsen 2009). Climate change and environmental degradation present emerging challenges that are reshaping forest resource utilization patterns worldwide (Schoene and Bernier 2012). Changing precipitation patterns, temperature regimes, and extreme weather events are altering forest ecosystem productivity and species composition, potentially affecting the availability and quality of forest resources upon which communities depend. Communities are developing various adaptation strategies, including diversifying their resource portfolios, modifying collection practices, and seeking alternative livelihood options (Sunderland et al. 2014).

Understanding these complex relationships between livelihoods and forest resources is essential for developing effective policies and interventions that can support both human well-being and forest conservation. This requires moving beyond simplistic narratives of forest degradation caused by local communities toward more nuanced analyses that recognize the sophisticated resource management systems developed by forest-dependent populations and the multiple external pressures that influence their practices. The following sections will explore these themes in greater detail, examining specific case studies, analyzing key drivers of change, and discussing policy implications for sustainable forest management that recognizes and supports the livelihood needs of forest-dependent communities while promoting conservation objectives.

7.2 Livelihood Dependencies and Forest Resource Utilization Patterns in West Bengal

West Bengal, characterized by its diverse physiographic regions from the sub-Himalayan Terai and Dooars in the north to the plateau fringe areas of the southwest is home to approximately 11,879 km^2 of forest cover, which constitutes around 13.38% of the state's total geographical area (FSI 2023). These forests are not only ecological lifelines but also socio-economic anchors for nearly 3 million forest-dependent people, primarily comprising Scheduled Tribes and Other Backward Classes (GoWB 2021). In districts such as Bankura, Purulia, Jhargram, and Paschim Medinipur, a significant share of the rural population depends on forests for basic subsistence needs fuelwood, fodder, thatching materials, and particularly Non-Timber Forest Products (NTFPs), which contribute 25–40% of household income among tribal families (Bisui and Shit 2023). NTFPs like *Shorea robusta* (sal leaves and seeds), *Madhuca indica* (mahua flowers), *Diospyros melanoxylon* (kendu leaves), tamarind, honey, lac, mushrooms, and medicinal plants (e.g., *Andrographis paniculata* and *Terminalia bellirica*) are harvested primarily during specific seasons (Kumar and Saikia 2020). These resources are intricately linked to indigenous ecological calendars, rituals, and women-led economies. For instance, in Jangalmahal (which includes parts of Bankura, Purulia, and Jhargram), over 70% of tribal women participate in NTFP collection and processing (Das et al. 2021). Their participation, however, is constrained by market access, price volatility, and policy limitations, as most of the trade is mediated through unorganized and exploitative channels. Despite initiatives like the Forest Rights Act, 2006, and Joint Forest Management (JFM), issues such as non-recognition of community forest rights, inadequate benefit sharing, and bureaucratic control continue to affect sustainable forest-livelihood integration (Xaxa 2014). A study by Bhattacharya and Basu (2018) found that only 30% of eligible households in West Bengal had received individual forest rights by 2017, and collective rights were nearly absent. Moreover, in many JFM committees, representation of women and marginalized groups remains symbolic, leading to inequitable decision-making and resource allocation (Das 2011). Forest degradation due to over-harvesting, invasive species, and shifting land use further endangers both ecological balance and livelihood security.

Recent remote sensing studies reveal notable forest fragmentation in the south-western districts between 1990 and 2020, with a 12–18% reduction in dense forest cover in Bankura and Purulia, largely due to human-induced pressures (Mandal and Chatterjee 2023). These changes have direct repercussions on livelihood stability. Field-based surveys indicate increasing seasonal migration of forest dwellers, particularly youth, due to declining resource availability and income from NTFPs. On the other hand, value-addition practices (e.g., sal plate making, lac cultivation, mahua fermentation) and women's Self Help Groups (SHGs) have shown promising results in enhancing income resilience, but require institutional support, infrastructure, and market linkages (Sinha et al. 2022). In sum, forest-based livelihood systems in West

Fig. 7.1 Symmetrical diagram of forest-based livelihood challenges and opportunities in West Bengal

Bengal are marked by spatial variability, gendered labor roles, and complex ecological dependencies (Fig. 7.1). A sustainable livelihood approach must therefore integrate community-led forest governance, legal rights recognition, climate-resilient harvesting practices, and geospatial monitoring of forest health. Geographers and policy planners must recognize the multifunctionality of forests not only as biodiversity hotspots, but also as living spaces of socio-economic survival especially for marginalized communities across the forest belts of West Bengal.

7.3 Leveraging Geospatial Technologies: GIS Tools and Techniques for Quantifying Livelihood Dependencies and Forest Utilization Patterns

The integration of Geographic Information Systems (GIS) provides indispensable analytical power for quantifying, visualizing, and understanding the complex spatial dimensions of livelihood dependencies on forests and associated resource utilization patterns (Fig. 7.2). Moving beyond traditional survey data alone, GIS enables researchers to geo-locate households, map resource collection areas, quantify spatial accessibility, analyze landscape changes over time, and model interactions between socio-economic factors and the forest environment. Key calculation tools and techniques include:

Spatial Buffering and Accessibility Analysis: Creating buffers around settlements or individual households (e.g., 1 km, 5 km) allows calculation of forest area *accessible* within specific travel-time or distance thresholds, a critical determinant of

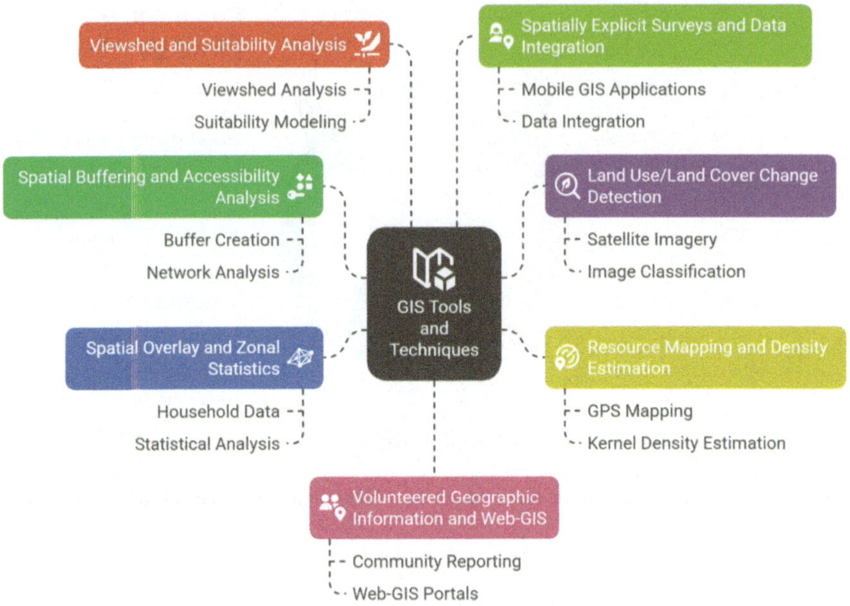

Fig. 7.2 Symmetrical diagram of GIS tools and techniques for forest dependency analysis

dependence intensity (Lele et al. 2010). Network analysis tools model travel paths along trails or rivers to derive more realistic travel times/costs to forest patches or specific resource locations (NTFP hotspots, water sources), revealing hidden spatial inequities in access (Fisher, 2004).

Land Use/Land Cover (LULC) Change Detection: Utilizing multi-temporal satellite imagery (Landsat, Sentinel) processed through GIS enables quantification of forest cover change (deforestation, degradation, regeneration) within specific zones relevant to communities (e.g., customary territories, Forest Rights Act [FRA] claim areas) (Roy et al. 2015). Techniques like image classification (supervised/unsupervised), NDVI trend analysis, and post-classification comparison provide metrics on forest loss/gain, directly linking to potential shifts in resource availability and utilization pressures.

Resource Mapping and Density Estimation: GPS-based participatory mapping integrates local knowledge by having community members geo-locate key resource collection sites (e.g., medicinal plants, fuelwood zones, fishing grounds) (Rambaldi et al. 2006). Kernel Density Estimation (KDE) within GIS then calculates the spatial density of these points, identifying core utilization zones and potential over-exploitation hotspots.

Spatial Overlay and Zonal Statistics: Overlaying household point locations (with survey data on income sources, asset wealth, caste/tribe, forest product use) with GIS layers of forest type, protected area boundaries, land tenure, infrastructure (roads), and terrain (slope, elevation) allows statistical analysis (e.g., zonal statistics, spatial regression) to identify *where* and *for whom* forest dependence is highest, and how utilization patterns correlate with biophysical and institutional factors (Das 2011).

Viewshed and Suitability Analysis: Viewshed analysis identifies forest areas *visible* from settlements, potentially influencing cultural value or monitoring ease. Suitability modeling (using Multi-Criteria Evaluation-MCE) combines factors like slope, soil, forest type, and distance to create maps predicting potential areas for specific NTFP collection or sustainable harvesting zones, informing management plans (Eastman 2015).

Spatially Explicit Surveys and Data Integration: Mobile GIS applications (like ODK Collect with GPS) enable surveyors to capture spatially referenced household data directly in the field, ensuring accurate linkage between socio-economic responses and geographic location. Integrating this point data with raster (forest cover, DEM) and vector (village boundaries, roads, rivers) layers within a GIS database is fundamental for multi-dimensional spatial analysis (Longley et al. 2015).

Volunteered Geographic Information (VGI) and Web-GIS: Platforms allowing communities to directly report resource sightings, collection areas, or conflicts via mobile apps contribute to VGI, enriching spatial datasets and fostering participatory monitoring (Goodchild 2007). Web-GIS portals can visualize spatial patterns of dependence and utilization for stakeholders and policymakers.

Critically, GIS analysis must be grounded in robust socio-economic data and local ecological knowledge. While GIS excels at quantifying spatial patterns, accessibility, and environmental change, understanding the *drivers* and *meanings* of dependence requires integrated mixed-methods approaches. Furthermore, ethical considerations regarding data privacy, community ownership of spatial information, and the potential for misuse in restricting access are paramount (Rambaldi et al. 2006). When applied thoughtfully, however, these GIS tools and techniques transform fragmented data into powerful spatial narratives, revealing hidden patterns and providing an evidence base for designing spatially targeted, equitable, and sustainable forest governance and livelihood interventions.

7.4 A Case Study on Livelihood Dependencies and Forest Resource Utilization Patterns in Binpur-II CD Block of Jhargram District, West Bengal

The Binpur-II Community Development block in Jhargram district exemplifies the complex forest-dependent livelihood patterns characteristic of West Bengal's Sal Forest regions (Fig. 7.3). This area, where Sal forests (*S. robusta*) dominate 60% of the landscape, demonstrates critical relationships between forest proximity, market accessibility, and household economic strategies (Sahoo et al. 2023). Forest dependency patterns reveal significant spatial variations across village classifications.

Households in High Forest-High Access (HFHA) zones generate Rs. 14,230 annually from forest resources, substantially higher than the Rs. 5077 earned by families in Low Forest-Low Access (LFLA) areas (Sahoo et al. 2023). This disparity highlights how both resource availability and market connectivity determine economic outcomes for forest-dependent populations. Non-Timber Forest Products (NTFPs) form the economic backbone, with 18 species supporting diverse livelihood activities. Sal leaves contribute 47% of NTFP income through plate manufacturing, while seasonal products like Mahua flowers provide concentrated income during March–May harvests (Bisui et al. 2021; Mukul et al. 2016). Gender dynamics significantly influence collection patterns, with women handling small-scale gathering of food and medicine while men engage in large-scale commercial activities. Regression analysis reveals that female-headed and larger households demonstrate significantly higher NTFP utilization rates ($p \leq 0.01$).

Concerning trends emerge from five-year temporal analysis showing resource depletion driven by climate variability, agricultural expansion, and unsustainable harvesting practices (Magry et al. 2022). Low-income households face particular vulnerability, deriving up to 43% of total income from forest resources, making them highly susceptible to environmental degradation (Vedeld et al. 2007). Market proximity critically influences economic viability, with Sal leaves commanding Rs. 200–220 per thousand plates in remote areas versus Rs. 230–250 in accessible locations, reflecting transportation costs and intermediary exploitation (Wunder et al. 2014). Socio-economic stratification reveals persistent poverty, with 32% earning below Rs. 2000 monthly.

Educational disparities compound inequalities, showing inverse relationships between forest proximity and literacy rates, potentially limiting livelihood diversification (Belcher et al. 2015). The demographic composition includes 30.7% Scheduled Tribes and 46.5% Other Backward Classes, though caste shows limited direct correlation with forest income levels while influencing access to alternative opportunities (Persha et al. 2011). Binpur-II's experience demonstrates both resilience and vulnerability of forest-based systems, emphasizing the need for integrated approaches combining conservation with livelihood security for India's 275 million forest-dependent people (Sukhdev et al. 2008). The Binpur-II Community Development block in Jhargram district, West Bengal, exemplifies India's forest-dependent rural landscapes where ecological resources intersect with socio-economic challenges. This administrative unit encompasses a distinctive Sal Forest (*S. robusta*) ecosystem constituting approximately 65% of the forested area, creating one of eastern India's most concentrated Sal zones (Sahoo et al. 2023; Bisui and Shit 2023). The remaining landscape comprises mixed plantations, scrub jungles, and bush formations reflecting natural succession and anthropogenic modifications (Bisui and Shit 2023).

This geographical setting provides an optimal framework for examining Non-Timber Forest Product (NTFP) extraction patterns and their impacts on rural livelihoods. The research adopts a spatially explicit approach investigating how forest proximity determines community dependency, recognizing that distance-decay relationships significantly influence resource accessibility (Shackleton and Pandey

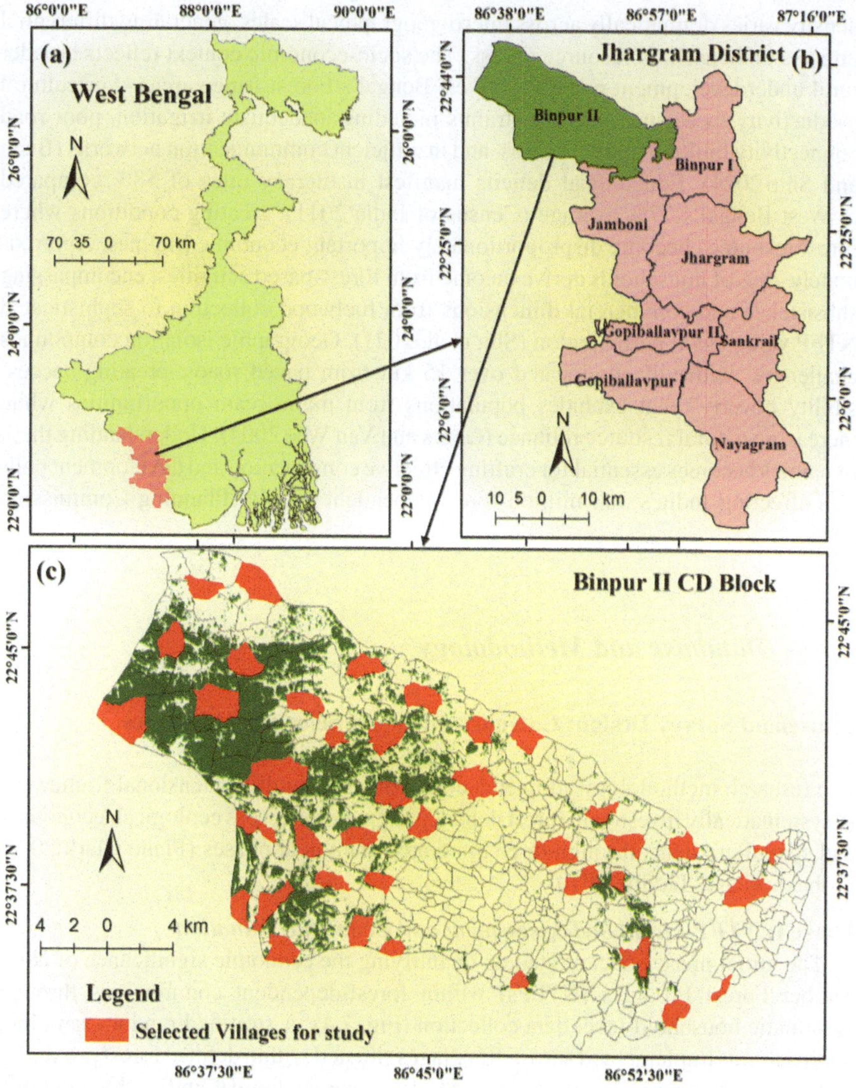

Fig. 7.3 Location map of the study area

2014). The analysis encompasses complex interactions between market accessibility and forest proximity, acknowledging that economic viability depends on integrating ecological and market factors (Belcher et al. 2005; Wunder et al. 2014). Binpur-II has experienced substantial ecological transformation over three decades, with deforestation accelerating since the 1990s. Urban encroachment has contributed to a 23% reduction in dense forest cover between 2000 and 2020 (Forest Survey of India, 2021). This fragmentation creates heterogeneous landscape mosaics where forest density varies dramatically across micro-geographical scales, generating differential impacts on household resource access. The socio-economic context reflects broader rural underdevelopment patterns in West Bengal's Forest fringe areas. Agricultural productivity faces structural constraints including inadequate irrigation, poor road connectivity limiting market access, and insufficient communication networks (Bisui and Shit 2024). Educational deficits manifest in literacy rates of 58% compared to West Bengal's 77% average (Census of India 2011), creating conditions where forest resources become disproportionately important economic lifelines. Approximately 78% of households derive income from forest-based activities, encompassing subsistence and commercial dimensions from fuelwood collection to sophisticated NTFP value-chain participation (Shit et al. 2021). Geographic isolation compounds challenges, with villages located over 15 km from paved roads, creating "accessibility poverty" that excludes populations from mainstream opportunities while increasing natural resource reliance (Geurs and Van Wee 2004). Understanding these dynamics becomes essential for crafting effective conservation and development policies affecting India's 300 million forest-dependent people (Planning Commission 2013).

7.4.1 Database and Methodology

Household Survey Design: Categorization and Selection of Villages

The research methodology adopted a comprehensive multi-dimensional framework to systematically investigate forest dependency patterns across ecological, economic, and spatial gradients through three integrated analytical phases (Plano Clark 2017; Tashakkori and Teddlie 2010).

Phase I: ***NTFP Livelihood Assessment and Economic Valuation***

The initial phase concentrated on quantifying the economic significance of Non-Timber Forest Products (NTFPs) within forest-dependent communities through systematic household-level data collection (Fig. 7.4). A stratified random sampling approach was implemented across 40 villages situated within defined forest proximity zones, ensuring representative coverage of diverse ecological and socio-economic conditions (Cochran 1977). The sampling framework incorporated 200 respondents selected through proportional allocation across demographic strata including caste categories, household sizes, and occupational groups, ensuring adequate population representation (Lohr 2021).

Primary data collection utilized structured questionnaires incorporating both quantitative economic indicators and qualitative livelihood assessment modules. The survey instrument captured temporal variations in NTFP collection patterns, seasonal income fluctuations, and household dependency ratios, following established methodologies in forest economics research (Angelsen et al. 2014; Sunderlin et al. 2005). Economic valuation employed market price assessments combined with

Fig. 7.4 Photos taken during the household field survey of the study area conducted through personal interviews and focus group discussions: **a** A group of researchers and local community members are seen engaged in an outdoor meeting seated on a mat. **b** This close-up view depicts a researcher in a focused conversation with local residents

shadow pricing techniques for subsistence consumption, enabling comprehensive quantification of both monetary and non-monetary forest contributions to household welfare (Vedeld et al. 2007).

Phase II: ***Spatial Analysis of Market-Forest Accessibility Dynamics***

The second phase employed Geographic Information System (GIS)-based spatial analysis to investigate relationships between market accessibility and forest proximity in determining livelihood outcomes (Fig. 7.5 and Table 7.1). High-resolution satellite imagery from Landsat-8 and Sentinel-2 platforms was processed using supervised classification algorithms to generate accurate forest cover maps with ground-truthing validation achieving 87% accuracy (Congalton and Green 2019). Road network analysis utilized OpenStreetMap data supplemented with field-verified GPS coordinates to construct comprehensive accessibility surfaces using cost-distance modeling techniques (Borkowska and Pokonieczny 2022).

Villages were systematically categorized into a four-class typology based on binary classifications of forest proximity (high/low) and road accessibility (high/low). Classification thresholds were established using k-means clustering analysis, with forest proximity measured as Euclidean distance to forest edges and road accessibility quantified through travel-time calculations to nearest paved roads (Fortin and Dale 2009). Systematic random sampling within each accessibility class yielded 40 villages distributed proportionally across categories, with 200 households selected for detailed livelihood surveys.

Phase III: ***Forest Proximity Gradient Analysis***

The final phase investigated forest dependency patterns along spatial gradients from forest interior to exterior locations. Village selection employed purposive sampling based on proximity classifications: inner villages (0–100 m from forest core), fringe villages (100–300 m from forest edge), and outer villages (> 500 m

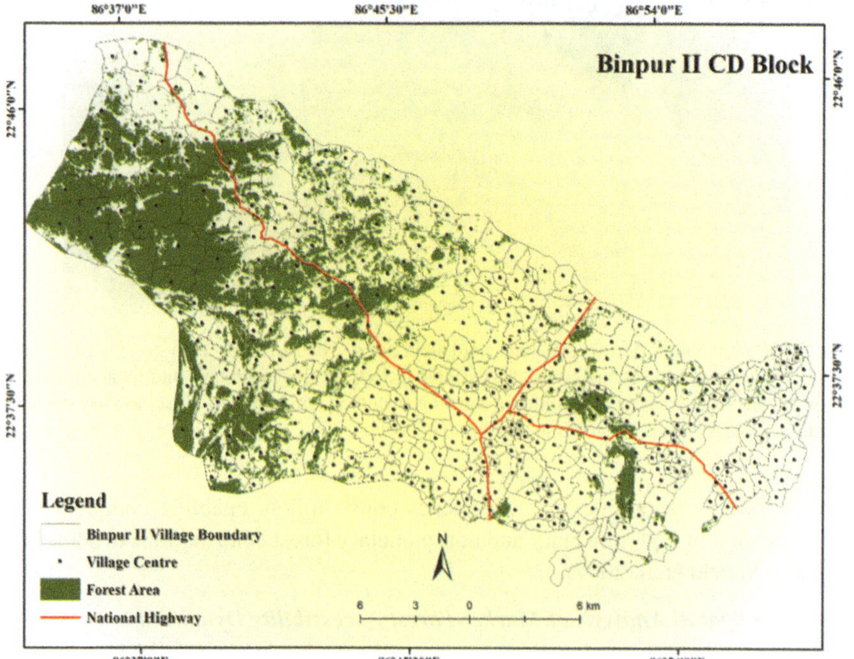

Fig. 7.5 Spatial distribution of village categories in the Binpur-II CD Block according to forest proximity and access to markets

Table 7.1 List of surveyed villages categorized by different village classes

HFHRA	HFLRA	LFHRA	LFLRA
Amjharna	Majugara	Astajuri	Harda
Jabla	Singaduba	Baramsol	Bara Salbani
Dangardiha	Amlasol	Orgonda	Mahulbani
Girulia	Kashijora	Bhula Bheda	Bodadihi
Beniadihi	Odolchua	Bidiri	Loyada
Birmadal	Barasol	Lagadari	Pachapani
Bamni	Kankrijharna	Tulsibani	Dhabani
Kaladabar	Shankhabhanga	Pukhuria	Shyampur
Kalapathar	Thakuranpahari	Dakaisol	Kapatkata
Gochgiri	Baishtampur	Dumur Kunda	Dholbhanga

N.B. *HFHRA* high forest and high road access, *HFLRA* high forest and low road access, *LFHRA* low forest and high road access, *LFLRA* low forest and low road access

from forest boundary) (Fig. 7.6 and Table 7.2). Distance thresholds were established through analysis of NTFP collection patterns from preliminary fieldwork and existing literature on optimal foraging distances (Hegde and Lingappa 1996; Pattanayak and Sills 2001). High-resolution satellite imagery combined with GPS-based ground-truthing enabled precise village boundary delineation and forest-edge mapping. Door-to-door household surveys employed random sampling with sample sizes calculated using power analysis to ensure adequate statistical power for detecting meaningful differences across proximity gradients (Cohen 1988).

The tri-phasic approach enabled comprehensive triangulation of findings across different analytical scales and methodological perspectives, enhancing internal and external validity (Patton 2014). Data quality assurance incorporated multiple validation procedures including inter-enumerator reliability assessments, GPS-verified location accuracy, and cross-validation through community focus groups. Statistical analysis employed hierarchical modeling to account for nested data structures and spatial autocorrelation effects (Gelman and Hill 2007), providing robust empirical foundations for understanding forest-livelihood relationships while addressing limitations in previous studies.

The research combined three distinct methods to build a complete picture of how forest dependency affects these communities economically and socially (Fig. 7.7). First, it looked at non-timber forest products. Second, it examined how easily people could reach markets and how close they lived to forests. Third, it explored what living near forests actually meant for residents. This careful, evidence-based approach strengthened the study's credibility and offered meaningful findings about how forests support people's lives in rural areas.

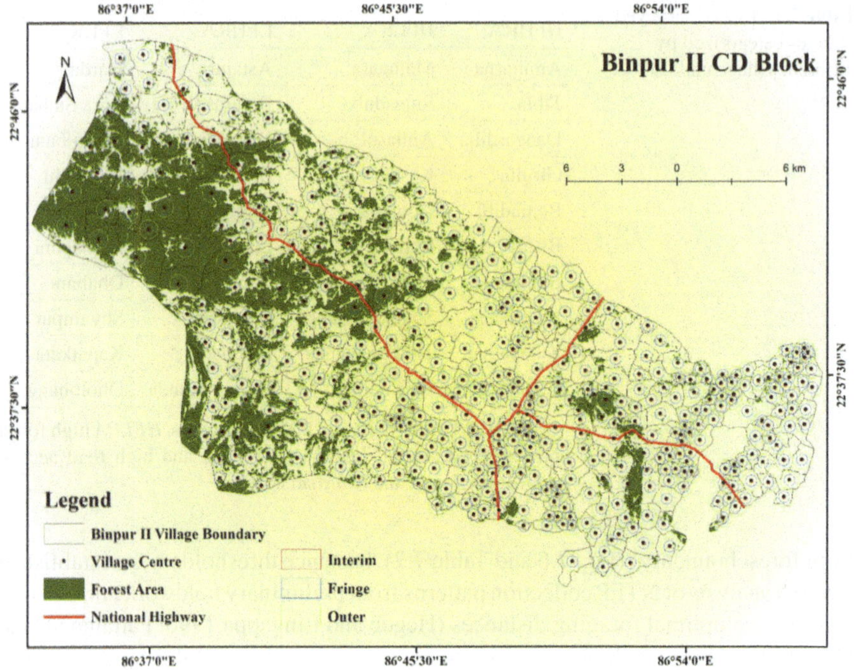

Fig. 7.6 Spatial classification of villages by forest proximity

Focus Group Discussions

This research implemented 25 Focus Group Discussions (FGDs) as a systematic participatory methodology to comprehensively examine the intricate relationships between forest resources, household economies, and livelihood strategies within forest-fringe communities (Krueger and Casey 2015). The FGDs served as critical platforms for fostering rich dialogue with village communities and multiple stakeholders, enabling contextual exploration of socio-economic dynamics that quantitative surveys alone cannot adequately capture (Liamputtong 2011). Participants were purposively selected through stratified random sampling to ensure demographic heterogeneity and representativeness across key population segments (Patton 2014). The sampling framework deliberately included community members from varying socio-economic strata, local governance representatives, forest-dependent households, and other relevant stakeholders, ensuring marginalized groups were adequately represented (Bernard 2017). Each discussion group comprised 7–10 carefully selected participants a composition size supported by methodological literature as optimal for generating substantive interaction while maintaining manageability (Nyumba et al. 2018).

Sessions lasting 15–25 min were conducted in familiar, accessible public venues including Primary Schools and Integrated Child Development Services (ICDS)

Table 7.2 List of surveyed villages categorized by different village classes based on forest proximity

S.No.	Surveyed villages	Village levels
01	Amjharna	Interim
02	Jabla	
03	Dangardiha	
04	Girulia	
05	Beniadihi	
06	Birmadal	
07	Bamni	
08	Kaladabar	
09	Kalapathar	
10	Gochgiri	
11	Majugara	
12	Singaduba	
13	Amlasol	
14	Kashijora	
15	Odolchua	Fringe
16	Barasol	
17	Kankrijharna	
18	Shankhabhanga	
19	Thakuranpahari	
20	Baishtampur	
21	Astajuri	
22	Baramsol	
23	Orgonda	
24	Bhula Bheda	
25	Bidiri	
26	Lagadari	
27	Tulsibani	
28	Pukhuria	Outer
29	Dakaisol	
30	Dumur Kunda	
31	Harda	
32	Bara Salbani	
33	Mahulbani	
34	Bodadihi	
35	Loyada	
36	Pachapani	
37	Dhabani	
38	Shyampur	

(continued)

Table 7.2 (continued)

S.No.	Surveyed villages	Village levels
39	Kapatkata	
40	Dholbhanga	

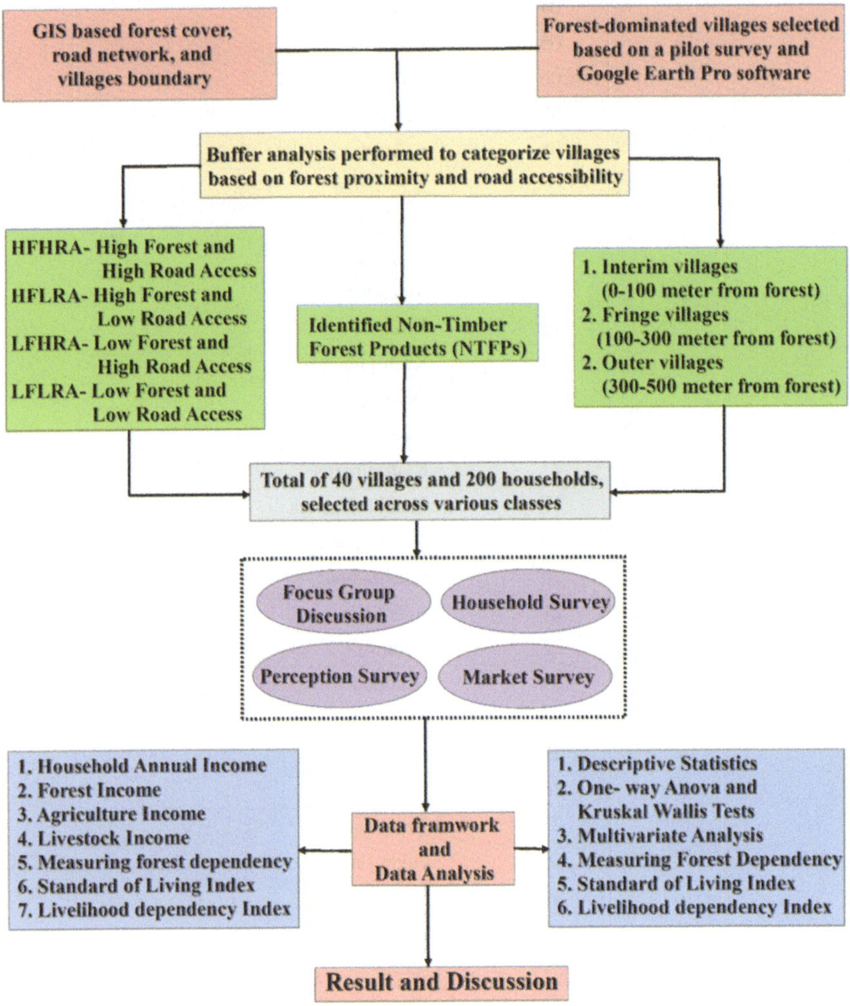

Fig. 7.7 Methodological framework

centers, deliberately chosen to reduce power asymmetries and create comfortable environments conducive to open dialogue (Peek and Fothergill 2009). Each session was facilitated by Gram Panchayat Members who possessed cultural competency and established community trust, enhancing participant comfort and response authenticity (Stewart and Shamdasani 2014). Discussions were structured around three interconnected thematic domains: First, participants articulated the fundamental role of Non-Timber Forest Products (NTFPs) (Fig. 7.8) in sustaining local economies, providing experiential accounts of how medicinal plants, wild foods, fuelwood, and minor forest produce contribute to household provisioning and market-oriented activities (Angelsen et al. 2014). Second, they traced historical transformations in household income generation, examining temporal evolution from traditional forest-based subsistence to diversified income strategies and the structural barriers encountered (Ellis 2000). Third, discussants explored underlying motivations for livelihood diversification, unpacking the complex interplay of push factors (resource scarcity, declining forest productivity) and pull factors (market opportunities, skill development initiatives) shaping household decisions (Sunderland et al. 2014).

The qualitative data generated through FGDs provided essential contextual depth that complemented quantitative household surveys, facilitating methodological triangulation and enhancing analytical robustness (Creswell and Plano Clark 2017). By integrating three analytical dimensions NTFP centrality in local economies, mediating influences of market accessibility and forest proximity, and broader socio-economic impacts of forest nearness this mixed-methods architecture provides a holistic understanding of structural and agential factors shaping economic resilience in forest-dependent communities (Scoones 2013), strengthening the validity, reliability, and transferability of research findings.

Data Analysis

This research employed a rigorous mixed-methods analytical framework to systematically interpret quantitative and qualitative data derived from household surveys and Focus Group Discussions (FGDs), enabling comprehensive triangulation and a deeper understanding of complex socio-economic phenomena (Creswell and Plano Clark 2017). The analytical strategy was deliberately structured around three interconnected dimensions to holistically examine the forest-livelihood nexus: first, the spatial dynamics of forest proximity and its influence on resource access patterns and household dependency levels (Angelsen et al. 2014); second, the economic significance and utilization patterns of Non-Timber Forest Products (NTFPs) in sustaining rural livelihoods and income portfolios (Shackleton and Pandey 2014); and third, the mediating role of market accessibility in shaping livelihood diversification opportunities and forest resource commercialization (Chamberlin and Jayne 2013). This tri-dimensional analytical approach facilitated a nuanced investigation of how geographical, ecological, and economic factors interact to determine household vulnerability, adaptive capacity, and livelihood resilience in forest-dependent communities (Ellis 2000). By integrating spatial, product-specific, and market-oriented analyses, this methodology provides robust empirical foundations for understanding the multifaceted relationships between forest ecosystems and human welfare, thereby enhancing the validity and policy relevance of research findings (Babulo et al. 2008).

Fig. 7.8 Some photographs of the study area show various resource collection activities

Descriptive Statistical Analysis

Descriptive statistical techniques were systematically employed to synthesize demographic attributes and critical socio-economic variables, providing a baseline understanding of data structure and distribution patterns (Pallant 2020). Comprehensive analyses encompassed measures of central tendency (arithmetic means), frequency distributions, and percentages to elucidate household income profiles, forest resource dependency intensities, and multi-dimensional socio-economic characteristics (Agresti et al. 2009).

These foundational summaries facilitated data exploration, identification of distributional characteristics, and detection of anomalies influencing subsequent analyses (Tukey 1977). The framework systematically documented variability across household demographics, income diversification, asset ownership, and forest product collection frequencies, establishing empirical baselines for comparative analysis and enhancing the interpretability of multivariate findings (Bernard 2017).

Income Calculation

Total Household Income: Calculated as the total income generated from agricultural production, forest resource utilization, wage work, and additional economic activities.

$$E_h = \sum_{i=1}^{n} [s_i] \tag{7.1}$$

where E_h is the total income earned by a household (INR); i corresponds to an individual income source; s_i denotes the cumulative earnings from that source (INR); and n defines the upper limit representing the total number of income categories included.

Forest Income: Forest-based income was computed by summing the monetary returns from key Non-Timber Forest Products such as fuelwood, medicinal plants, sal leaves and seeds, honey, kendu leaves, and mahua flowers (Vedeld et al. 2007) (Table 7.3).

$$E_f = \sum_{i=1}^{n} [F_i P_i - (K_i)] \tag{7.2}$$

where E_f denotes the total forest income; F_i represents the quantity of forest product i collected (kg); P_i indicates its corresponding market price (INR); and K_i refers to the production cost associated with each forest product (Table 7.1).

Agricultural Income: Agricultural income was estimated based on total crop production and corresponding market prices (INR/kg).

$$Y_a = \sum_{i=1}^{n} [A_c.P_i] \tag{7.3}$$

where Y_a stands for agricultural income in Indian Rupees (INR); A_c indicates the total crop output measured in kilograms; and P_i represents the respective market price of crop i.

Table 7.3 Market price of forest products in different forest classes

NTFPs	Market price in different forest ranges (1 US$ = 88.07 INR) (Oct 15, 2025)		
	Outer	Fringe	Interim
Sal leaf (1000 plates)	2.61–2.83	2.49- 2.61	2.27–2.49
Sal seeds (kg)	0.11–0.13	0.11	0.09–0.11
Mahua (kg)	0.79–0.91	0.74–0.85	0.74–0.79
Mahua seed (kg)	0.91–0.96	0.85–0.91	0.79–0.91
Mushrooms (kg)	1.48–1.70	1.36–1.48	1.14–1.42
Atanglata (1bojha = 10 basket)	5.68	5.68	5.68
Kendu (1 pan = 1600 leaf)	0.68–0.74	0.62–0.68	0.62–0.68
Khejure leaf (1 bojha = 30 Broom)	2.84–3.41	2.27–2.84	2.27–2.50
Behera (kg)	0.11–0.17	0.11–0.17	0.09–0.11
Fuelwood (kg)	0.06–0.07	0.05–0.06	0.05–0.06

Livestock Income: Livestock income was estimated based on the number of animals owned and their corresponding market prices (INR/kg).

$$Y_l = \sum_{i=n}^{n} [N_i.P_i - (K_i)] + \sum_{i=1}^{n} [Q_i P_i - (K_i)] \tag{7.4}$$

where Y_l represents the total livestock income; N_i denotes the number of livestock animals; P_i is the market price per unit; K_i indicates the maintenance or rearing cost; and Q_i refers to the quantity of product obtained from livestock i (kg).

Statistical Analysis

To examine variations in income sources and dependency patterns, inferential statistical methods were applied, including t-tests and analysis of variance (ANOVA). These tests compared villages categorized by their spatial characteristics (interim, fringe, and outer zones) and transportation connectivity (high versus low road access), where significant differences emerged. Tukey's post-hoc test identified which specific village groups differed from one another. Additionally, multiple regression modeling evaluated how various socio-economic and demographic variables influenced the extent of forest resource use among households.

Forest Dependency

Forest dependency was quantified by dividing total forest-derived income by overall household income.

$$RFI = \frac{TotalForestIncome(TFI)}{TotalHouseholdIncome(THI)} \quad (7.5)$$

where RFI represents the relative forest income, TFI denotes total income derived from forest resources, and THI indicates the overall household income.

Standard of Living Index (SLI)

The standard of living index was constructed using the methodology developed by the International Institute for Population Sciences and ORC Macro (2000) for the National Family Health Survey (NFHS). This approach incorporates 27 indicators encompassing housing quality, farm equipment ownership, utility access (including water supply, electricity, and cooking fuel), and possession of household consumer goods. Each indicator received a predetermined weight, and the final index score was calculated by aggregating these weighted values.

Livelihood Dependency Index (LDI)

To quantify the role of forest products in sustaining household livelihoods, a Livelihood Dependency Index (LDI) was employed. This index classifies households according to their reliance on forest resources (Schmerbeck and Fiener 2015) (Table 7.4).

$$LDI = (nD_1 \times 1) + (nD_2 \times 0.75) + (nD_3 \times 0.50) + (nD_4 \times 0.25) + (nD_5 \times 0) \quad (7.6)$$

where n represents the total number of surveyed households, whereas D_1, D_2, D_3, D_4, D_5 and D_n denote the various categories of forest products considered (Table 7.1).

Thematic Analysis: Data collected through focus group discussions were transcribed verbatim and subjected to thematic analysis. A coding framework was applied to identify emergent themes concerning economic activities, income portfolio diversification, and adaptive responses to livelihood challenges. Through this rigorous methodological approach, we systematically examined the complex relationships between geographical distance from forests, patterns of NTFP collection and use, and accessibility to commercial markets. This analysis provided valuable insights into how these interconnected variables influence the economic well-being of rural households within the research sites.

Table 7.4 Significance of forest resources for livelihood maintenance

Code	Description
D1	Can't do without
D2	Important
D3	Not so important
D4	Not important at all
D5	Don't know/no answer

7.4.2 Results and Discussion

Livelihood Contributions and Consequences of Non-timber Forest Product (NTFP) Extraction

Socio-economic and Demographic Characteristics

The socio-economic profile of the 400 sampled households reveals significant demographic and economic stratification patterns. Male household heads constitute a slight majority (56.71%), reflecting traditional patriarchal structures, though substantial female headship (43.25%) indicates evolving gender dynamics in rural governance (Table 7.5). Educational attainment remains predominantly concentrated at medium levels (58.50%), with alarmingly minimal higher education penetration (0.01%), highlighting critical human capital development gaps that constrain livelihood diversification opportunities. Caste composition demonstrates pronounced social heterogeneity, with Other Backward Classes forming the largest segment (42%), followed by Scheduled Tribes (31%). This distribution underscores the region's socially marginalized character, necessitating targeted developmental interventions. Household size predominantly ranges between one and four members (55.25%), suggesting nuclear family transitions from traditional joint family systems, potentially driven by economic migration and resource constraints.

The age structure concentrates heavily within the economically productive 35-49 years bracket (53.75%), representing optimal demographic dividend potential. However, subsistence income patterns reveal pronounced economic vulnerability, with a mean annual income of ₹80,876.17 heavily dependent on forest resources (₹35,467.72) and wage labor (₹20,983.55). This dual dependency exposes households to ecological degradation risks and labor market fluctuations. Agriculture contributes marginally (₹5598.10), indicating either limited landholding or productivity constraints. Livestock income (₹11,913.89) provides supplementary support, while negligible business income (₹937.41) reflects restricted entrepreneurial engagement. The substantial forest income component emphasizes the critical ecosystem-livelihood nexus, warranting sustainable forest management policies that balance conservation imperatives with community welfare requirements.

Livelihood Structure and Income Composition

The income portfolio composition reveals pronounced dependence on non-timber forest products (NTFPs), which constitute nearly half (47%) of total household income, establishing forest ecosystems as the primary economic lifeline for rural communities (Fig. 7.9). This overwhelming reliance underscores the intricate relationship between ecological resources and livelihood sustainability, positioning forest conservation as inseparable from poverty alleviation strategies. Wage labor emerges as the secondary income source (22%), reflecting limited asset ownership and compelling households toward daily labor markets, thereby exposing them to seasonal employment fluctuations and income volatility.

Table 7.5 Details of the socio-economic and demographic characteristics ($N = 400$)

Household class	Attributes	Frequency	Minimum	Maximum	Mean ± SD
Household head	M	227 (56.71%)	1	2	1.38 ± 0.48
	F	173 (43.25%)			
Education	LOWER	161 (40.25%)	1	3	1.51 ± 0.53
	MEDIUM	234 (58.50%)			
	HIGHER	05 (0.01%)			
Cast	SC	63 (16.75%)	1	4	2.44 ± 0.87
	ST	124 (31%)			
	OBC	168 (42%)			
	GEN	45 (13.5%)			
Sex	M	228 (57%)	1	2	1.39 ± 0.48
	F	172 (43%)			
Household size	01–4	221 (55.25)	2.00	13.00	1.53 ± 0.45
	05–9	132 (33%)			
	10–14	47 (11.75%)			
Age	20–34	101 (25.25%)	25	65	43.63 ± 7.92
	35–49	215 (53.75)			
	50–64	73 (18.25%)			
	Above 64	11 (2.75%)			
Subsistence Income	30–50 k	16 (4%)	45,100.0	130,500.0	80,876.17 ± 13,282.55
	50–70 k	79 (19.75%)			
	70–90 k	193 (48.25%)			
	Above 90 k	112 (28%)			
Forest income			9600.00	69,400.00	35,467.72 ± 8821.81
Livestock income			0	28,244	11,913.89 ± 7856.19

(continued)

Table 7.5 (continued)

Household class	Attributes	Frequency	Minimum	Maximum	Mean ± SD
Wages income			10,000	35,000	20,983.55 ± 5332.61
Agriculture income			0	15,000	5598.1 ± 5082.78
Business income			0	30,000	937.41 ± 4478.51
Others income (including remittance)			0	50,000	3519 ± 2536.53
Total income			44,200	126,700	80,787.7 ± 13,278.8

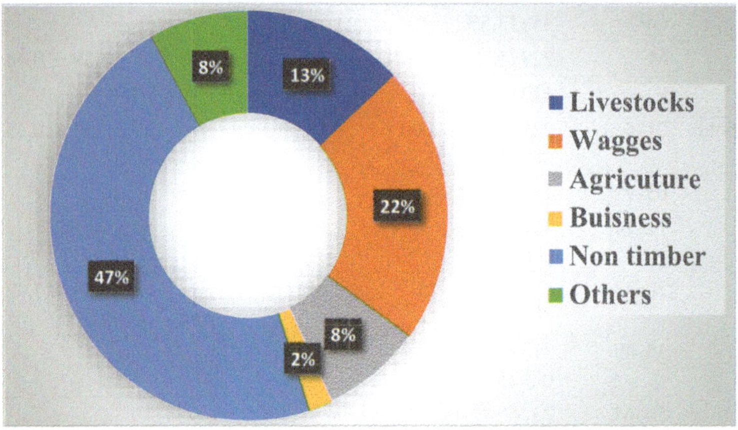

Fig. 7.9 Percentage contributions of livestock, wages, and agriculture, business, and non-timber forest products incomes to the total household income

Livestock-based earnings contribute moderately (13%), suggesting complementary pastoral activities that provide risk-buffering mechanisms against forest resource variability and agricultural uncertainties. However, agricultural income remains surprisingly marginal (8%), indicating either constrained land access, degraded soil productivity, or inadequate irrigation infrastructure that limits cultivation potential. This agricultural underperformance perpetuates forest dependency cycles, as households compensate through intensified NTFP extraction.

Business entrepreneurship contributes negligibly (2%), revealing structural barriers including insufficient capital access, limited market linkages, and restricted skill diversification opportunities. Miscellaneous income sources, including remittances (8%), provide supplementary support but remain inadequate for transformative economic mobility. This income structure epitomizes a forest-dependent economy characterized by livelihood vulnerability and limited economic diversification. The predominance of extractive and labor-intensive activities, coupled with minimal value-addition enterprises, suggests an urgent necessity for integrated development approaches that enhance forest productivity sustainably while simultaneously creating alternative income pathways through skill development, microenterprise promotion, and improved market access infrastructure.

Agricultural and commercial entrepreneurship exhibited comparatively modest contributions to aggregate household income streams. NTFP harvesting and livestock-based revenues demonstrated statistically significant superiority ($P < 0.05$) over cultivation-derived earnings and other resource categories. This economic pattern establishes forest resource extraction and labor market participation as primary income determinants within the study population, while agricultural commodity sales and animal husbandry functioned as supplementary revenue sources. Business-oriented income generation remained conspicuously constrained in household economic strategies.

Household Income Derived from Non-timber Forest Products (NTFPs) Over the Past Five Years

The multifunctional utilization of non-timber forest products demonstrates remarkable diversity in household consumption patterns, with construction materials dominating usage at 34%, followed by fodder (29%), food (20%), and cash income generation (17%) (Fig. 7.10). This distribution reveals that forest resources primarily serve subsistence needs rather than market-oriented exploitation, underscoring their foundational role in meeting basic survival requirements. The substantial allocation toward construction materials reflects traditional housing practices and limited access to alternative building resources, while the significant fodder contribution sustains livestock populations that provide supplementary nutrition and income.

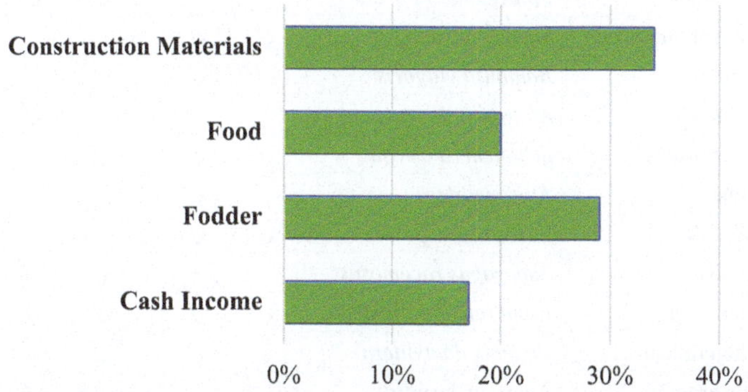

Fig. 7.10 Various uses of non-timber forest resources

The identification of eighteen distinct NTFP species highlights exceptional biodiversity utilization spanning medicinal plants (*A. paniculata, Asparagus racemosus*), nutritional resources (*Madhuca longifolia, Phyllanthus emblica*), and economically valuable products (*D. melanoxylon* leaves, *S. robusta* components) (Fig. 7.6). Temporal harvesting patterns reveal strategic resource collection concentrated during February through June, coinciding with pre-monsoon and early monsoon periods when agricultural activities remain minimal and households face acute income deficits. This seasonal dependency creates critical livelihood vulnerabilities during resource scarcity periods.

The predominance of indigenous species like Sal, Mahua, Kendu, and Bamboo emphasizes deep ecological knowledge systems embedded within local communities. However, the heavy reliance on forest biomass for construction (34%) and sustenance needs raises sustainability concerns regarding extraction pressures on forest ecosystems. The modest cash income proportion (17%) suggests either limited market integration, inadequate value-chain development, or preferential subsistence consumption over commercialization. This utilization pattern necessitates balanced management frameworks that ensure ecological regeneration while supporting community welfare through sustainable harvesting protocols and market linkage enhancement (Table 7.6).

Table 7.6 Important NTFPs resources in the study region

S.No	NTFPs	Scientific name	Harvesting time				
			Feb	March	Apr	May	June
1	Amla	*Phyllanthus emblica*	√	√			
2	Bel	*Aegle marmelos*			√	√	√
3	Kendu leaf	*Diospyros melanoxylon*			√	√	
4	Kendu fruit	*Diospyros melanoxylon*			√	√	√
5	Mahua Flower	*Madhuca longifolia*		√	√	√	
6	Mahua seeds	*Madhuca longifolia*				√	√
7	Dudhilata	*Finlaysonia obovata*			√	√	√
8	Sal leaf and Seeds	*Shorea robusta*	√	√	√	√	√
9	Bamboo	*Bambusa vulgaris*	√	√	√		
10	Kalmegh	*Andrographis paniculata*			√	√	√
11	Kusum seeds	*Schleichera oleosa*	√	√	√		
12	Honey	*Apis mellifera*	√	√	√	√	√
13	Palas flower	*Butea monosperma*	√	√			
14	Satabari	*Asparagus racemosus*			√	√	√
15	Atanglata	*Combretum decandrum*	√	√	√	√	√
16	Khejure leaf	*Phoenix dactylifera*	√	√	√	√	√
17	Mushrooms	*Agaricus bisporus*			√	√	√
18	Bahera	*Terminalia bellirica*			√	√	√

Contribution of Non-timber Forest Products (NTFPs) to Household Income

The disaggregated analysis of NTFP contributions reveals pronounced concentration within the household income portfolio, with Sal (*S. robusta*) leaf and seed collection singularly dominating at 47%, establishing it as the quintessential economic resource for forest-dependent communities. This overwhelming reliance reflects both the species' ecological abundance in the study region and established market mechanisms for Sal-based products, particularly leaf plates utilized extensively across traditional dining practices. The substantial market demand, coupled with year-round availability, positions Sal as an indispensable livelihood anchor.

Secondary contributors include Kendu (*D. melanoxylon*) leaf (7%), Mahua (*M. longifolia*) flower (5%), and several products contributing 3–4% each including mushrooms, Sal seeds, Amla (*P. emblica*), honey, and bamboo (Fig. 7.11). This tiered contribution pattern illustrates livelihood diversification strategies wherein households combine high-volume, low-value products (Sal leaves) with specialized, seasonally available resources. Kendu leaf collection, traditionally linked to beedi (indigenous cigarette) manufacturing, provides concentrated seasonal income despite regulatory constraints.

Numerous species contribute marginally (1–2%), including Bahera (*T. bellirica*), Bel (*Aegle marmelos*), Kendu fruit, and Kusum seeds (*Schleichera oleosa*), reflecting either limited market demand, restricted harvesting rights, or minor ecological availability. This income fragmentation across eighteen species demonstrates sophisticated ecological resource utilization but simultaneously exposes vulnerability to species-specific market fluctuations and regulatory interventions. The extreme dependence on Sal products warrants concern regarding over-harvesting pressures and necessitates rotational harvesting protocols. Geographic concentration of single-species dependency limits household resilience against ecological disturbances,

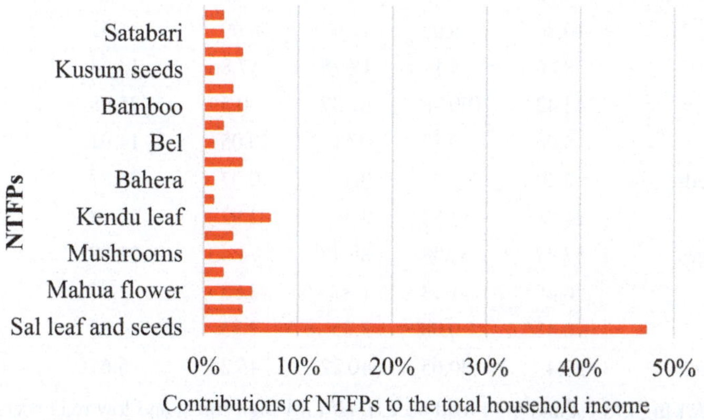

Fig. 7.11 Percentage contributions of NTFPs to the total household income

market disruptions, or policy restrictions, emphasizing urgent requirements for value-chain enhancement and equitable benefit-sharing mechanisms.

Forest-Based Livelihood Strategies: Market Accessibility and Forest Proximity Economic Contributions of Forest Products to Household Incomes by Access Class

The spatial differentiation of household income sources across access classes reveals compelling evidence of location-dependent livelihood strategies shaped by forest proximity and road connectivity. High Forest-High Road Access (HFHRA) communities demonstrate significantly elevated per-capita earnings from forest-dependent resources, particularly fuelwood (US \$68.18), livestock (US \$242.42), and leaf plates (US \$24.85), compared to Low Forest-Low Road Access (LFLRA) villages ($P<0.05$) (Table 7.7). These spatial gradient underscores how environmental endowment and infrastructural accessibility synergistically determine economic opportunities, with forest-proximate settlements capitalizing on both resource availability and market linkages.

Table 7.7 Annual per-capita income (US\$) from various sources by access class

Source of income	Village class wise annual per-capita income (US\$)				ANNOVA_F	*P*-value
	HFHRA	HFLRA	LFHRA	LFLRA		
Fuel wood	68.18	68.18	46.55	25.86	61.48	0.001
Fodder	28.86	30.7	2.74	1.65	149.23	0.003
Lac	0.01	0.01	0	0	5.36	0.005
Kendu leaf	3.72	0.81	0.66	0.27	25.11	0.001
Fruits	0.17	0.06	0.03	0.01	5.89	0.003
Medicine	0.16	0.25	0.03	0.06	5.38	0.001
Mahua	7.67	0.68	0.1	0.03	65.71	0.001
Mushroom	2.12	1.9	0.67	0.23	43.12	0.001
Bamboo	10.26	6.06	11.6	8.69	1.52	0.265
Agro-forest	8.26	8.1	18.75	17.86	13.64	0.001
Livestock	242.42	190.38	61.57	60.59	75.98	0.002
Sal seeds	5.06	0.52	0.11	0.05	14.91	0.005
Mahua seeds	2.29	0.76	0.3	0.07	35.89	0.007
Plates	24.85	13.54	8.49	2.92	66.18	0.001
Labor wages	57.97	52.48	88.48	74.43	21.89	0.001
Trade	0.45	1.93	1.87	2.98	2.87	0.061
Remittance	0.43	0.98	3.25	2.85	4.95	0.002
Agriculture	37.4	30.65	40.22	45.24	5.67	0.002

N.B. *HFHRA* high forest and high road access, *HFLRA* high forest and low road access, *LFHRA* low forest and high road access, *LFLRA* low forest and low road access

Conversely, LFHRA and LFLRA communities exhibit compensatory livelihood adaptations, deriving substantially higher incomes from wage labor (US \$88.48 and US \$74.43, respectively) and agriculture (US \$40.22 and US \$45.24), suggesting deliberate shifts toward non-forest economic activities where forest resources remain scarce. The statistically significant variations ($P < 0.001$) in specialized NTFPs Mahua flowers (US \$7.67 in HFHRA versus US\$0.03 in LFLRA), mushrooms, and Kendu leaves highlight how ecological proximity enables exploitation of perishable, location-specific forest commodities that require immediate processing or swift market access.

Remarkably, bamboo income exhibits spatial homogeneity across access classes ($P = 0.265$), indicating widespread cultivation or equitable distribution mechanisms transcending geographical constraints. The progressive decline in fodder income from HFHRA (US \$30.70) to LFLRA (US \$1.65) directly correlates with diminished livestock holdings in forest-deprived areas. Road accessibility emerges as critical, with HFHRA communities leveraging superior connectivity for high-value forest product commercialization, while isolated HFLRA villages remain constrained despite forest abundance. This spatial analysis illuminates how geographical positioning fundamentally structures livelihood portfolios, necessitating spatially differentiated development interventions.

Forest Dependence by Different Access Class

The per-capita forest product income distribution across access classes demonstrates striking spatial inequalities, with absolute forest income (AFI) declining sharply from HFHRA villages (US \$161.67 ± 50.58) to LFLRA settlements (US \$57.68 ± 34.08), representing a nearly threefold differential. This dramatic gradient underscores how geographical positioning combining forest availability and infrastructural connectivity fundamentally determines resource access and economic extraction capacity. The Kruskal–Wallis test results reveal statistically significant variations across multiple forest products ($P < 0.05$), confirming that spatial location constitutes a critical determinant of livelihood opportunities rather than random distribution patterns.

Paradoxically, relative forest income (RFI) exhibits inverse relationships with absolute earnings, where low-income households demonstrate highest forest dependence (0.49 ± 0.13) despite generating minimal absolute returns (US \$57.45 ± 27.28) (Table 7.8). This phenomenon epitomizes the "poverty-environment trap," wherein economically marginalized communities rely disproportionately on common-pool forest resources as safety-net mechanisms, yet their limited asset base, restricted market access, and constrained bargaining power prevent translation of resource dependence into substantial income generation. Medium and high-income groups show progressively declining RFI (0.37 and 0.34, respectively) while achieving substantially higher AFI (US \$111.16 and US \$158.27), indicating successful livelihood diversification beyond forest extraction.

Product-specific analysis reveals spatial concentration patterns: fuelwood remains universally significant across access classes (0.43–0.62), reflecting fundamental energy requirements, while specialized products like Mahua (0.05 in HFHRA versus

Table 7.8 Annual per-capita income (US$) from various forest products in different access classes

Products	Village class wise annual per-capita income (US$)				Total	Kruskal–Wallis test		
	HFHRA	HFLRA	LFHRA	LFLRA		ANNOVA_F	df	*P*-value
Lack	0	0	0	0	0	0	0	0
Fuel	0.5	0.62	0.46	0.43	2.01	9.21	3	0
Fodder	0.21	0.27	0.03	0.03	0.54	100.01	3	0
Kundu	0.03	0.01	0.01	0.01	0.06	3.42	3	0.017
Fruits	0	0	0	0	0	0	0	0
Medicinal Plant	0	0	0	0	0	0	0	0
Mahua	0.05	0.01	0	0	0.06	88.15	3	0.002
Bamboo	0.06	0.04	0.16	0.11	0.37	4.91	3	0.01
Agro-forest	0.05	0.06	0.25	0.49	0.85	62.89	3	0.01
Mushroom	0.02	0.02	0.01	0	0.05	23.94	3	0.01
Plate	0.17	0.12	0.1	0.05	0.44	34.51	3	0.02
Sal Seeds	0.03	0	0	0	0.03	16.31	3	0
Mahua Seeds	0.01	0.01	0	0	0.02	25.74	3	0

N.B. *HFHRA* high forest and high road access, *HFLRA* high forest and low road access, *LFHRA* low forest and high road access, *LFLRA* low forest and low road access

0.00 in LFLRA) and leaf plates (0.17 vs. 0.05) demonstrate pronounced accessibility-dependent variations (Table 7.9). Interestingly, agroforestry contributions increase inversely with forest access (0.05 in HFHRA to 0.49 in LFLRA), suggesting deliberate cultivation strategies compensating for natural forest scarcity. The uniform RFI (0.37) across HFHRA, HFLRA, and LFHRA classes, despite divergent absolute incomes, indicates that forest resources maintain proportional economic significance regardless of total household earnings, until extreme resource scarcity (LFLRA) forces structural livelihood transformations. These findings necessitate equity-focused forest governance frameworks addressing both spatial marginalization and income-based vulnerabilities.

Forest Proximity and Household Dependency

Standard of Living Index in Different Village Classes

The standard of living index reveals a compelling inverse relationship between forest proximity and household welfare, challenging conventional assumptions about natural resource access and prosperity (Fig. 7.12). Interim villages, despite maximum

Table 7.9 Forest income by access class and income quartile: absolute and relative measures

Types	Income categories	
	AFI Mean ± SD	RFI Mean ± SD
HFHRA	161.67 ± 50.58	0.37 ± 0.08
HFLRA	129.66 ± 31.79	0.37 ± 0.06
LFHRA	90.02 ± 30.78	0.37 ± 0.10
LFLRA	57.68 ± 34.08	0.26 ± 0.11
Total	109.76	0.34
Low income group	57.45 ± 27.28	0.49 ± 0.13
Medium income group	111.16 ± 30.27	0.37 ± 0.08
High income group	158.27 ± 47.02	0.34 ± 0.07
Total	109.76	0.34

forest resource availability, exhibit the highest concentration of low-living-standard households (25 out of 126, 19.8%) and minimal high-standard representation (7.1%).

Conversely, outer villages geographically remote from forests demonstrate zero low-standard households and the highest proportion achieving elevated living standards (26 out of 124, 21%). This paradoxical pattern suggests forest dependence may perpetuate poverty cycles through limited economic diversification, vulnerability to seasonal income fluctuations, and restricted integration with broader market economies. The income distribution visualization corroborates these welfare disparities, showing that interim villages derive approximately 60–70% of total earnings from forest resources, while outer settlements depend predominantly on agriculture (50-60%) with negligible forest contributions. Fringe villages occupy intermediate positions both geographically and economically, displaying balanced forest-agriculture portfolios. The overwhelming prevalence of medium-living-standard

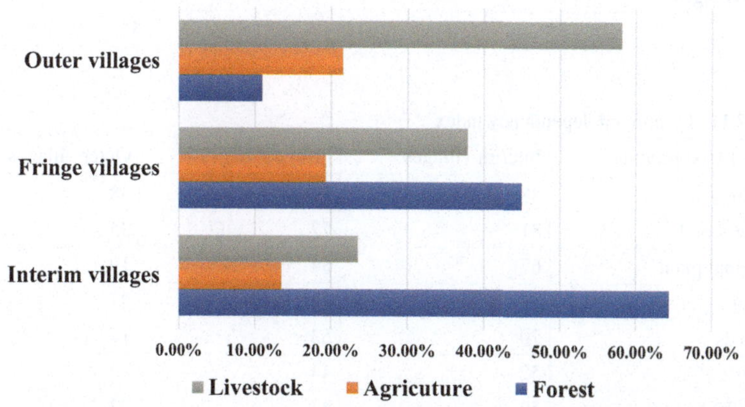

Fig. 7.12 Distribution of income sources across village classes

households across all categories (82.9%) indicates widespread economic stagnation transcending spatial variations (Table 7.10). These findings underscore that sustainable development requires transformative interventions beyond mere resource access, including value-chain enhancement, skill diversification, and infrastructural improvements that enable forest-dependent communities to translate ecological proximity into genuine prosperity rather than subsistence-level survival.

Livelihood Dependency Index Among Different Village Classes

The livelihood dependency indices exhibit pronounced distance-decay patterns, systematically declining from interim to outer villages across all forest product categories. Sal leaf extraction demonstrates maximum dependency in interim settlements (94), declining progressively through fringe (78) to outer villages (27), representing a 71.3% reduction attributable to both ecological scarcity and accessibility constraints (Table 7.11). Similarly, fuelwood dependency shows substantial spatial variation (88–38), though remaining relatively significant even in remote settlements, reflecting its fundamental role in household energy requirements where alternative fuel sources remain economically prohibitive or infrastructurally unavailable.

Specialized non-timber products display steeper dependency gradients: Mahua flowers decline from 49 to 17, Sal seeds from 70 to 17, and medicinal plants from 67 to 19, indicating these resources require immediate forest proximity for viable exploitation. The consistently high interim village indices across timber (81), medicinal plants

Table 7.10 Standard of living index

Village category	Standard of living index (weighted score)		
	Low (0–20)	Medium (20–40)	High (40–67)
Interim villages	25	92	9
Fringe villages	4	121	0
Outer villages	0	98	26
Total	29	311	35

Table 7.11 Livelihood dependency index

Forest-based product	Interim villages	Fringe villages	Outer villages
Fuelwood	88	79	38
Timber wood	81	77	35
Medicinal plant	67	54	19
Sal leaf	94	78	27
Sal seeds	70	53	17
Honey	52	44	18
Kendu leaf	59	43	22
Mahua flower	49	25	17
Mahua seeds	45	26	13

(67), and Kendu leaves (59) underscore comprehensive forest resource integration within daily subsistence activities. Conversely, outer villages demonstrate uniformly low dependencies (13–38 range), signaling near-complete livelihood detachment from forest ecosystems and structural economic reorientation toward alternative income sources. This spatial dependency configuration highlights how geographical positioning determines not merely resource access quantities but fundamentally shapes household survival strategies, with forest-adjacent communities exhibiting existential reliance while distant settlements develop compensatory non-forest-based economic portfolios necessitated by ecological remoteness.

Livelihood Activities of Households Across Different Village Classes

The mean income distribution across village classes reveals dramatic livelihood restructuring driven by forest accessibility. Interim villages generate substantially higher forest income (US $637.98) compared to fringe (US $371.77) and outer villages (US $141.93), representing a 78% decline across the spatial gradient (Table 7.12). This forest income concentration is overwhelmingly dominated by Mahua flower collection (US $629.30) in interim settlements, constituting 98.6% of their forest earnings a perilous single-commodity dependence exposing households to market volatility and seasonal income shocks. Fuelwood emerges as the most spatially resilient forest product (US $303.89 to US $107.46), maintaining economic significance even in remote locations due to universal energy requirements.

Conversely, agricultural income exhibits inverse spatial patterns, escalating from US $149.90 in interim villages to US $260.55 in outer settlements, a 73.8% increase reflecting deliberate cultivation intensification compensating for forest resource inaccessibility. Specialized NTFPs including medicinal plants, Sal seeds, honey, Kendu leaves, and Mahua seeds generate zero income in outer villages, indicating complete market exclusion from these forest-dependent value chains. Remarkably,

Table 7.12 Sources of income (US$) among village classes

Source of income	Interim villages (mean)	Fringe villages (mean)	Outer villages (mean)
Timber wood	156.79	95.91	30.2
Fuelwood	303.89	188.2	107.46
Medicinal plant	11.68	3.39	0
Sal leaf	173.83	66.11	4.23
Sal seeds	14.44	7.73	0
Honey	2.78	2.32	0
Kendu leaf	34.84	3.22	0
Mahua flower	629.3	1.95	0
Mahua seeds	7.17	3.04	0
Forest income	637.98	371.77	141.93
Agricultural income	149.9	193.62	260.55
Total income	805.68	611.56	603.29

total household income converges between fringe and outer villages (approximately US $611-603), suggesting multiple viable livelihood pathways achieving comparable economic outcomes through fundamentally different resource portfolios forest extraction versus agricultural production underscoring spatial adaptation strategies necessitated by geographical positioning.

7.4.3 Conclusion

The comprehensive analysis of livelihood dependencies and forest resource utilization patterns in Binpur-II CD block reveals a complex socio-ecological landscape where spatial positioning fundamentally determines economic trajectories and survival strategies. The empirical evidence demonstrates that forest-dependent communities navigate intricate trade-offs between resource accessibility and market integration, with households proximate to forests deriving substantial income from Non-Timber Forest Products particularly Sal leaves and Mahua flowers while simultaneously experiencing lower overall living standards compared to their geographically distant counterparts. This paradoxical relationship underscores that mere ecological proximity does not guarantee prosperity; rather, it often perpetuates subsistence-level dependencies that constrain livelihood diversification and economic mobility. The pronounced spatial gradients in forest dependency indices, coupled with inverse agricultural income patterns, illuminate adaptive livelihood strategies wherein remote communities deliberately intensify cultivation activities to compensate for forest resource inaccessibility. Gender dimensions, caste stratification, and educational deficits further compound these vulnerabilities, creating multi-layered marginalization that transcends simple resource availability considerations. The overwhelming concentration of forest income within single commodities, particularly Mahua flowers, contributing 98.6% of forest earnings in interim villages, exposes households to severe seasonal income volatility and market fluctuations. Moving forward, sustainable forest governance must transcend conservation-centric approaches to embrace spatially differentiated interventions that recognize diverse livelihood contexts strengthening value chains for forest-proximate communities while simultaneously enhancing agricultural productivity and market linkages for remote settlements. Only through such integrated, equity-focused frameworks can policy effectively balance ecological conservation imperatives with the legitimate welfare aspirations of forest-dependent populations across West Bengal's diverse physiographic regions.

References

Agrawal A, Angelsen A (2009) Using community forest management to achieve REDD+ goals. Realising REDD+: national strategy and policy options 1:201–212

Agarwal B (2015) The power of numbers in gender dynamics: illustrations from community forestry groups. J Peasant Stud 42(1):1–20
Agresti A, Finlay B, Porcu M (2009) Statistica per le scienze sociali. Pearson, London, UK
Angelsen A, Jagger P, Babigumira R, Belcher B, Hogarth NJ, Bauch S et al (2014) Environmental income and rural livelihoods: a global-comparative analysis. World Dev 64:S12–S28
Awazi NP (2025) Climate resilience in forest-dependent communities in the developing world: what implications for livelihoods and sustainability?. In: Building resilience: climate change and livelihoods in the global south. Springer Nature Switzerland, Cham, pp 27–54
Babulo B, Muys B, Nega F, Tollens E, Nyssen J, Deckers J, Mathijs E (2008) Household livelihood strategies and forest dependence in the highlands of Tigray, Northern Ethiopia. Agric Syst 98(2):147–155
Bank W (2016) World Bank Group Forest Action Plan FY16–20
Belcher B, Ruíz-Pérez M, Achdiawan R (2005) Global patterns and trends in the use and management of commercial NTFPs: implications for livelihoods and conservation. World Dev 33(9):1435–1452
Belcher B, Achdiawan R, Dewi S (2015) Forest-based livelihoods strategies conditioned by market remoteness and forest proximity in Jharkhand, India. World Dev 66:269–279
Berkes F, Colding J, Folke C (2000) Rediscovery of traditional ecological knowledge as adaptive management. Ecol Appl 10(5):1251–1262
Bernard HR (2017) Research methods in anthropology: qualitative and quantitative approaches. Bloomsbury Publishing PLC
Bhattacharya R, Basu P (2018) Pollinator limitation and crop production: experimental observations on few economically important vegetable crops in West Bengal, India. In: Proceedings of the zoological society, vol 71, No 1. Springer India, New Delhi, pp 88–91
Bisui S, Roy S, Sengupta D, Bhunia GS, Shit PK (2021) Assessment of ecosystem services values in response to land use/land cover change in tropical forest. In: Forest resources resilience and conflicts. Elsevier, pp 435–447
Bisui S, Shit PK (2023) Assessing the dependence of livelihoods in rural communities on tropical forests: insights from the Midnapore Forest Division in West Bengal, India. Socio-Ecol Pract Res 5(4):423–437
Bisui S, Shit PK (2024) Assessing the role of forest resources in improving rural livelihoods in West Bengal of India. RegSustain 5(2):100141
Borkowska S, Pokonieczny K (2022) Analysis of OpenStreetMap data quality for selected counties in Poland in terms of sustainable development. Sustainability 14(7):3728
Census of India (2011) Provisional population totals. Government of India, Ministry of Home Affairs. Retrieved from https://censusindia.gov.in
Chamberlin J, Jayne TS (2013) Unpacking the meaning of 'market access': evidence from rural Kenya. World Dev 41:245–264
Cochran WG (1977) Sampling techniques. Wiley
Cohen J (1988) Set correlation and contingency tables. Appl Psychol Meas 12(4):425–434
Colfer CJP (2013) The gender box: a framework for analysing gender roles in forest management, vol 82. CIFOR
Commission P (2013) Twelfth five year plan (2012–2017): three volume set, vol 1. SAGE Publications Pvt. Limited, India
Congalton RG, Green K (2019) Assessing the accuracy of remotely sensed data: principles and practices. CRC Press
Creswell JW, Clark VLP (2017) Designing and conducting mixed methods research. Sage publications
Das N (2011) Women's dependence on forest and participation in forestry: A case study of joint forest management programme in West Bengal. J Econ 17(1):67–89
Das D, Ghosh P, Das AA (2021) Non-timber forest produces (NTFPs) and livelihood security of people in West Bengal. In: Forest resources resilience and conflicts. Elsevier, pp 227–237

Eastman JR (2015) TerrSet: geospatial monitoring and modeling software. Clark University, Clark Labs

Ellis F (2000) Rural livelihoods and diversity in developing countries. Oxford University Press

Fisher M (2004) Household welfare and forest dependence in Southern Malawi. Environ Dev Econ 9(2):135–154

Forest Survey of India (2021) India State of Forest Report 2021. Ministry of Environment, Forest and Climate Change, Government of India. https://fsi.nic.in/isfr2021/

Fortin MJ, Dale MR (2009) Spatial autocorrelation in ecological studies: a legacy of solutions and myths. Geogr Anal 41(4):392–397

FSI (2023) India State of Forest Report 2023. Forest Survey of India, Ministry of Environment, Forest and Climate Change, Government of India

Gelman A, Hill J (2007) Data analysis using regression and multilevel/hierarchical models. Cambridge University Press

Geurs KT, Van Wee B (2004) Accessibility evaluation of land-use and transport strategies: review and research directions. J Transp Geogr 12(2):127–140

Goodchild MF (2007) Citizens as sensors: the world of volunteered geography. GeoJournal 69(4):211–221

GoWB (2021) Statistical handbook of West Bengal 2021. Bureau of Applied Economics and Statistics, Government of West Bengal

Hegde RS, Lingappa VR (1996) Sequence-specific alteration of the ribosome–membrane junction exposes nascent secretory proteins to the cytosol. Cell 85(2):217–228

Janse G, Ottitsch A (2005) Factors influencing the role of non-wood forest products and services. Forest Policy Econ 7(3):309–319

Krueger RA, Casey MA (2015) Focus group interviewing. Handbook of practical program evaluation, pp 506–534

Kumar R, Saikia P (2020) Forest resources of Jharkhand, Eastern India: socio-economic and bio-ecological perspectives. Socio-economic and eco-biological dimensions in resource use and conservation: strategies for sustainability. Springer International Publishing, Cham, pp 61–101

Lele S, Wilshusen P, Brockington D, Seidler R, Bawa K (2010) Beyond exclusion: alternative approaches to biodiversity conservation in the developing tropics. Curr Opin Environ Sustain 2(1–2):94–100

Liamputtong P (2011) Focus group methodology: principle and practice

Lohr SL (2021) Sampling: design and analysis. Chapman and Hall/CRC

Longley PA, Goodchild MF, Maguire DJ, Rhind DW (2015) Geographic information science and systems. Wiley

Magry MA, Cahill D, Rookes J, Narula SA (2022) An integrated value chain analysis of non-timber forest products: a case of Jharkhand State of India. Small-Scale Forestry 21(4):621–645

Mandal M, Chatterjee ND (2023) Geo-spatial analysis of forest landscape for wildlife management. Springer

Moran EF, Ostrom E (eds) (2005) Seeing the forest and the trees: human-environment interactions in forest ecosystems. Mit Press

Mukul SA, Rashid AM, Uddin MB, Khan NA (2016) Role of non-timber forest products in sustaining forest-based livelihoods and rural households' resilience capacity in and around protected area: a Bangladesh study. J Environ Planning Manage 59(4):628–642

Nyumba OT, Wilson K, Derrick CJ, Mukherjee N (2018) The use of focus group discussion methodology: insights from two decades of application in conservation. Methods Ecol Evol 9(1):20–32

Office of the Registrar General & Census Commissioner, India (2011) Census of India 2011. Government of India. https://censusindia.gov.in/

Pallant J (2020) Ebook: SPSS surival manual: a step by step guide to data analysis using IBM SPSS. McGraw-Hill Education (UK)

Pattanayak SK, Sills EO (2001) Do tropical forests provide natural insurance? The microeconomics of non-timber forest product collection in the Brazilian Amazon. Land Econ 77(4):595–612

Patton MQ (2014) Qualitative research & evaluation methods: Integrating theory and practice. Sage publications
Paumgarten F, Shackleton CM (2011) The role of non-timber forest products in household coping strategies in South Africa: the influence of household wealth and gender. Popul Environ 33(1):108–131
Pavri F, Deshmukh S (2003) Institutional efficacy in resource management: temporally congruent embeddedness for forest systems of western India. Geoforum 34(1):71–84
Peek L, Fothergill A (2009) Using focus groups: Lessons from studying daycare centers, 9/11, and Hurricane Katrina. Qual Res 9(1):31–59
Persha L, Agrawal A, Chhatre A (2011) Social and ecological synergy: local rulemaking, forest livelihoods, and biodiversity conservation. Science 331(6024):1606–1608
Plano Clark VL (2017) Mixed methods research. J Posit Psychol 12(3):305–306
Rambaldi G, Kyem PAK, McCall M, Weiner D (2006) Participatory spatial information management and communication in developing countries. Electron J Inf Syst Develop Countries 25(1):1–9
Roy PS, Behera MD, Murthy MSR, Roy A, Singh S, Kushwaha SPS et al (2015) New vegetation type map of India prepared using satellite remote sensing: Comparison with global vegetation maps and utilities. Int J Appl Earth Obs Geoinf 39:142–159
Sahoo KP, Roy A, Mandal MH, Yasmin B, Bhattacharjee S, Siddique G (2023) Appraisal of coexistence and interdependence of forest and tribes in Jhargram District of West Bengal, India using SWOT–AHP analysis. GeoJ 88(2):1493–1513
Schmerbeck J, Fiener P (2015) Wildfires, ecosystem services, and biodiversity in tropical dry forest in India. Environ Manage 56(2):355–372
Schoene DH, Bernier PY (2012) Adapting forestry and forests to climate change: a challenge to change the paradigm. Forest Policy Econ 24:12–19
Scoones I (2013) Livelihoods perspectives and rural development. In: Critical perspectives in rural development studies. Routledge, pp 159–184
Shackleton CM, Pandey AK (2014) Positioning non-timber forest products on the development agenda. Forest Policy Econ 38:1–7
Shackleton CM, Shackleton SE, Buiten E, Bird N (2007) The importance of dry woodlands and forests in rural livelihoods and poverty alleviation in South Africa. Forest Policy Econ 9(5):558–577
Shit PK, Pourghasemi HR, Adhikary PP, Bhunia GS, Sati VP (eds) (2021) Forest resources resilience and conflicts. Elsevier
Sinha AK, Thakur R, Khatri A (eds) (2022) Social impact assessment in India: learning from the field. SAGE Publishing India
Stewart DW, Shamdasani PN (2014) Focus groups: theory and practice. Sage publications
Sukhdev P, Wittmer H, Schröter-Schlaack C, Neßhöver C, Bishop J, Brink PT et al (2008) The economics of ecosystems and biodiversity. na
Sunderlin WD, Angelsen A, Belcher B, Burgers P, Nasi R, Santoso L, Wunder S (2005) Livelihoods, forests, and conservation in developing countries: an overview. World Dev 33(9):1383–1402
Sunderland T, Achdiawan R, Angelsen A, Babigumira R, Ickowitz A, Paumgarten F et al (2014) Challenging perceptions about men, women, and forest product use: a global comparative study. World Dev 64:S56–S66
Tashakkori A, Teddlie C (eds) (2010) Sage handbook of mixed methods in social & behavioral research. Sage
Tukey JW (1977) Exploratory data analysis. Reading/Addison-Wesley
Vedeld P, Angelsen A, Bojö J, Sjaastad E, Berg GK (2007) Forest environmental incomes and the rural poor. Forest Policy Econ 9(7):869–879
Wunder S, Angelsen A, Belcher B (2014) Forests, livelihoods, and conservation: broadening the empirical base. World Dev 64:S1–S11
Xaxa V (2014) Report on the high-level committee on socio-economic, health and educational status of tribal communities of India

Chapter 8
Joint Forest Management: Success Stories and Implementation Challenges

Abstract This chapter critically examines Joint Forest Management (JFM) implementation in West Bengal's Arabari Forest Range, the pioneering site that catalyzed participatory forestry governance across South Asia since 1972. Employing mixed-methods research encompassing 140 households across eight forest-fringe settlements, the study integrates household surveys, focus group discussions, GIS-based spatial analysis, and ecosystem service valuation to assess JFM's socio-ecological outcomes over five decades. Results demonstrate remarkable ecological restoration, with forest canopy density recovering from below 20% to over 60%, and ecosystem service values increasing from US$6.02 million to US$7.17 million (1990–2023). Livelihood assessments reveal 94% household access to non-timber forest products and 76% receiving forest royalty benefits, validating resource-dependent poverty alleviation claims. However, critical governance deficits fundamentally undermine JFM's democratic aspirations. Power structure analysis exposes systematic marginalization, with women comprising merely 5.51% of committee membership and catastrophically declining to 0.88% in grievance mechanisms, while Scheduled Castes and Scheduled Tribes represent only 19.93% and 18.07%, respectively, despite profound forest dependencies. Community satisfaction surveys reveal stark contradictions: while 80% approve ecological restoration, only 46% perceive genuine empowerment through usufruct rights, with 54% expressing dissatisfaction regarding socio-political objectives. The Livelihood Dependency Index highlights differentiated resource hierarchies, with fuelwood (LDI 73) and Sal leaves (LDI 69) dominating subsistence portfolios. Findings underscore that ecological success does not automatically translate into social justice, revealing how JFM committees institutionalize pre-existing patriarchal and caste-based hierarchies rather than democratizing resource governance, thereby necessitating fundamental institutional reforms prioritizing substantive participation, transparent benefit-sharing, and genuine authority devolution to achieve equitable socio-ecological futures.

Keywords Joint forest management · Arabari forest range · Livelihood dependency index · Scheduled castes · Scheduled Tribes

P. K. Shit and M. Hasanuzzaman, *Forest Landscape Dynamics in West Bengal*, Advances in Asian Human-Environmental Research,
https://doi.org/10.1007/978-3-032-23700-2_8

8.1 Introduction

Joint Forest Management (JFM) emerged in the late twentieth century as a paradigm-shifting approach to forest governance, fundamentally challenging the legacy of state-centric, exclusionary forestry models prevalent across much of the Global South (Agarwal 2019). Born from grassroots innovations, notably the Arabari experiment in West Bengal, India, during the 1970s, JFM formally institutionalized the principle that the long-term health of forest ecosystems and the well-being of adjacent communities are inextricably linked and best served through collaborative partnerships between state forest departments (SFDs) and local village communities (Poffenberger and McGean 1996; Bisui et al. 2023). At its core, JFM represents a contractual arrangement where communities, organized into Forest Protection Committees (FPCs) or similar institutions, assume responsibilities for forest protection, management, and restrained resource extraction. In return, they are granted legally defined usufruct rights to specific non-timber forest products (NTFPs), a share of timber revenue upon maturity, and sometimes access to minor timber or fuelwood (Agrawal and Ostrom 2001; Shit et al. 2021). This model promised a triple win: halting forest degradation through local stewardship, alleviating rural poverty by recognizing forest-dependent livelihoods, and empowering communities through decentralized decision-making.

The proliferation of JFM, particularly across South and Southeast Asia (e.g., India, Nepal, and the Philippines), represents one of the most significant institutional experiments in participatory natural resource management (Bhattacharya et al. 2010). Its success stories, documented in diverse ecological and socio-political contexts, offer compelling evidence of its potential. Empirical studies have shown that well-implemented JFM can lead to measurable improvements in forest condition, including increased canopy cover, biodiversity recovery, and reduced incidence of fire and illicit felling, particularly in degraded forests where community vigilance proves effective (Watson et al. 2000; Behera and Engel 2006). Crucially, JFM has demonstrably enhanced livelihood security for millions by securing access to critical NTFPs, generating supplemental income through wage labor in plantation and protection activities, and providing a share of timber revenues, thereby reducing vulnerability and diversifying income sources (Lele et al. 2010). Furthermore, it has fostered social capital by strengthening village institutions, empowering marginalized groups (though unevenly), and creating platforms for negotiation between communities and the state (Agrawal and Gupta 2005).

However, JFM is far from a panacea. Its implementation has been fraught with persistent and multifaceted challenges that often undermine its theoretical promise. A critical issue is the persistent power asymmetry between SFDs and communities. SFDs frequently retain control over key decisions (species selection, harvesting schedules, fund allocation), treating FPCs as labor forces rather than genuine partners, reflecting a lack of genuine devolution of authority (Shit et al. 2021). Elite capture within villages is rampant, where benefits (leadership positions, resource access, funds) are disproportionately cornered by wealthier, higher caste, or male members, marginalizing the poor, women, Scheduled Tribes (STs), and Scheduled Castes (SCs) whose dependence on forests is often highest (Agarwal 2009). Ambiguous or insecure tenurial rights plague many JFM agreements, leaving communities vulnerable to eviction or sudden changes in policy, undermining long-term investment in forest management (Springate-Baginski and Blaikie 2007). Inadequate benefit-sharing mechanisms, bureaucratic delays in revenue distribution, and restrictions on harvesting high-value products often render the economic incentives insufficient to motivate sustained community effort (Lele et al. 2010). Finally, JFM struggles with scaling complexities effective models in one ecological or socio-cultural setting may fail in another, and institutional capacity for supporting thousands of FPCs is often lacking (Agrawal and Chhatre 2006). This sub-chapter delves into the intricate tapestry of JFM, critically examining its celebrated successes alongside its entrenched implementation challenges. It explores the diverse institutional forms JFM takes across regions, analyzes the factors contributing to its effectiveness or failure (including ecological context, market linkages, local institutional strength, and political will), and assesses its long-term sustainability as a model for reconciling conservation goals with the rights and needs of forest-dependent people in an era of increasing environmental stress and competing land-use demands.

8.2 Joint Forest Management: Success Stories and Implementation Challenges in West Bengal

West Bengal holds a unique position in the annals of forest governance as the birthplace of Joint Forest Management (JFM), with the pioneering Arabari experiment initiated in the Midnapore district in 1972 serving as the prototype for participatory forestry models globally (Shit et al. 2021) (Fig. 8.1). Faced with severe degradation of Sal (*Shorea robusta*) forests and escalating conflicts, the Forest Department forged agreements with local communities, forming Forest Protection Committees (FPCs) tasked with guarding forests against illicit felling and grazing. In return, communities received exclusive usufruct rights to non-timber forest products (NTFPs) and a significant share (typically 25-50%) of net revenue from the final timber harvest (Murali et al. 2003). This model, formally adopted statewide through government resolutions in 1989 and 1990, expanded rapidly, covering vast areas of degraded forests and involving millions of households, particularly Scheduled Tribes (STs) and Scheduled Castes (SCs) (Bisui et al. 2023). Documented success stories highlight measurable ecological and socio-economic gains: significant improvements in forest cover and density have been recorded in many JFM areas, particularly in the southwestern districts (Purulia, Bankura, West Midnapore), transforming degraded scrub into productive coppice Sal stands and mixed forests through natural regeneration and limited plantation (Malhotra and Deb 1993; Watson et al. 2000). Concurrently, JFM provided crucial livelihood benefits by securing access to NTFPs like Kendu leaf (*Diospyros melanoxylon*) a major state-monopolized product providing seasonal cash income to collectors fuelwood, fodder, and medicinal plants, alongside generating wage employment through forest protection and plantation works (Ghate 2003). Furthermore, the institutional framework fostered the development of social capital at the village level through FPCs, creating platforms for collective action and negotiation, albeit often dominated by local elites (Edmunds and Wollenberg 2013).

However, the implementation of JFM in West Bengal has been fraught with deep-seated challenges and contradictions that have often undermined its equity and sustainability objectives (Fig. 8.2). A fundamental issue remains the persistent power asymmetry between the Forest Department and FPCs. Despite the rhetoric of partnership, the Forest Department frequently retains de facto control over key management decisions (species selection, harvesting schedules, fund allocation), treating FPCs more as labor contractors than genuine co-managers, reflecting a lack of meaningful devolution of authority (Springate-Baginski and Blaikie 2007). This power imbalance is exacerbated by rampant elite capture within FPCs. Village elites, often from dominant castes or classes, disproportionately control FPC leadership positions and decision-making, marginalizing the voices and interests of the poor, women, landless, and tribal communities who are often the most forest-dependent (Agarwal 2009).

Consequently, the distribution of tangible benefits remains highly skewed. While wage labor from plantation work is relatively widespread, access to lucrative NTFP collection contracts (like Kendu leaf) and, crucially, the highly anticipated timber revenue shares have often been delayed, mismanaged, or captured by elites, failing to provide adequate long-term incentives for many marginalized members (Lele et al. 2010). The tenurial security offered by JFM has also been ambiguous and contested,

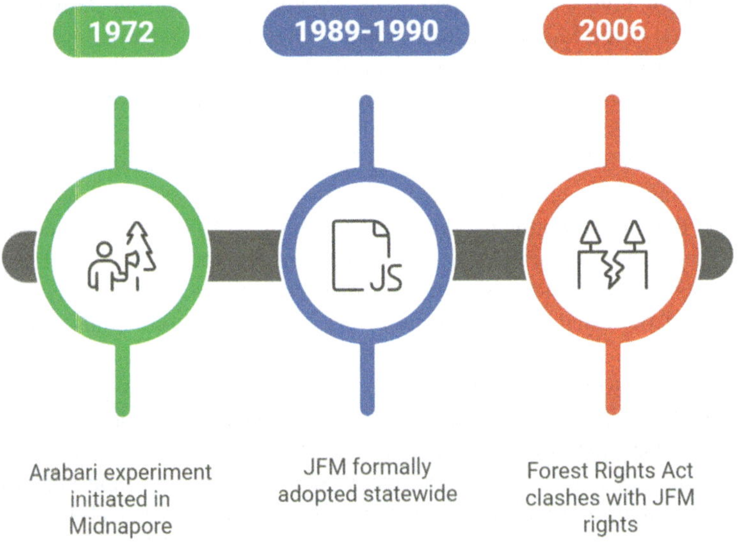

Fig. 8.1 Symmetrical diagram of West Bengal's journey in joint forest management

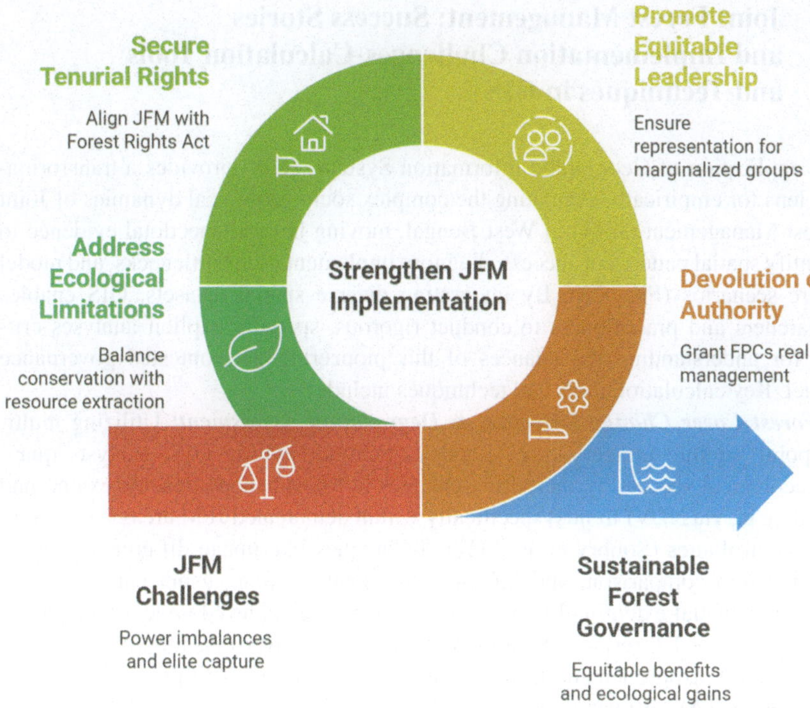

Fig. 8.2 Symmetrical diagram of enhancing joint forest management in West Bengal

particularly in relation to the Forest Rights Act (FRA) 2006. JFM rights, seen by the Forest Department as privileges granted under departmental control, often clash with FRA's recognition of community forest resource (CFR) rights as inherent entitlements, creating confusion and insecurity for communities (Larson 2010). Additionally, ecological limitations are evident: the model has been most successful in regenerating degraded Sal forests but faces difficulties in complex ecosystems like the Sundarbans mangroves, where resource extraction needs careful balancing with conservation, and where climate change impacts (sea-level rise, cyclones) add severe stress to both forests and JFM institutions (DasGupta and Shaw 2017). Thus, while West Bengal's JFM pioneered a new paradigm and achieved notable localized successes, its implementation reveals enduring institutional weaknesses, power imbalances, and unresolved tensions over rights and benefits, highlighting the gap between participatory ideals and on-ground realities in forest governance.

8.3 Joint Forest Management: Success Stories and Implementation Challenges-Calculation Tools and Techniques in GIS

The application of Geographic Information Systems (GIS) provides a transformative lens for empirically evaluating the complex socio-ecological dynamics of Joint Forest Management (JFM) in West Bengal, moving beyond anecdotal evidence to quantify spatial patterns of success, diagnose implementation bottlenecks, and model future scenarios (Fig. 8.3). By integrating diverse spatial datasets, GIS enables researchers and practitioners to conduct rigorous, spatially explicit analyses critical for understanding the nuances of this pioneering yet contested governance model. Key calculation tools and techniques include:

Forest Cover Change Detection & Degradation Assessment: Utilizing multi-temporal satellite imagery (e.g., Landsat, Sentinel) within GIS, analysts quantify ecological success by measuring changes in forest canopy density, extent, and health (e.g., via NDVI trends) specifically within demarcated JFM areas versus non-JFM control sites (Soubry et al. 2021). Techniques like image differencing, post-classification comparison, and fragmentation analysis (e.g., using patch metrics) reveal the spatial extent and intensity of regeneration achieved under JFM, particularly in the Sal forests of Southwest Bengal, or conversely, identify persistent degradation hotspots demanding intervention (Biswas 2010). This provides objective evidence of JFM's impact on forest condition.

Spatial Accessibility Modeling: GIS network analysis tools calculate realistic travel times/costs from villages to JFM forest boundaries and key resource zones

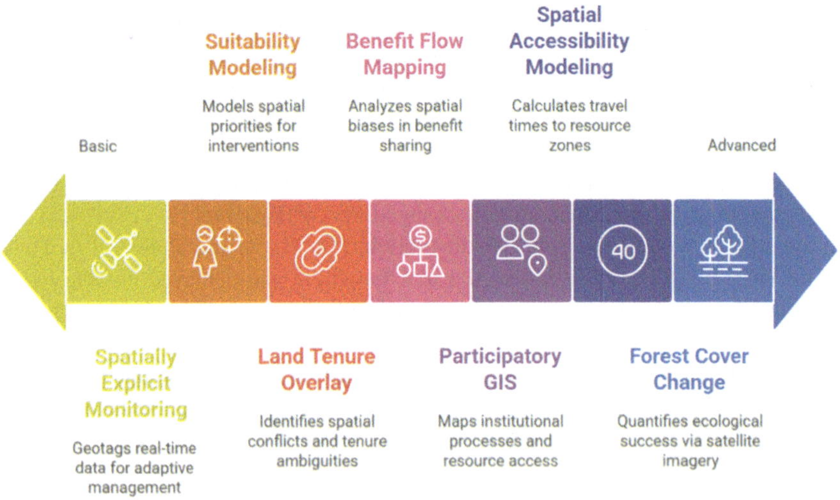

Fig. 8.3 Symmetrical diagram of GIS applications in joint forest management (JFM), ranging from basic to advanced

(NTFP collection areas, timber coupes) (Lele et al. 2010). Creating buffers or service areas based on these costs reveals spatial inequities in access identifying marginalized hamlets (often SC/ST dominated) located beyond viable distances, thus excluded from purported JFM benefits despite nominal FPC membership, a critical factor in understanding skewed benefit distribution (Fisher 2004; Das 2014).

Participatory GIS (PGIS) for Institutional Mapping & Equity Analysis: Integrating GPS-based participatory mapping with household surveys captures the spatial dimension of institutional processes and resource access. Communities map FPC boundaries, resource collection zones, illicit felling/grazing hotspots, and perceived areas of elite control. Overlaying this with socio-economic data (caste, gender, wealth) in GIS allows spatial statistical analysis (e.g., spatial regression, cluster analysis) to detect patterns of elite capture for instance, whether high-value Kendu leaf collection areas disproportionately abut elite households' lands (Bisui et al. 2023). Mapping women's specific collection routes and zones highlights gendered access limitations.

Benefit Flow Mapping and Zonal Statistics: Georeferencing data on timber revenue distribution, NTFP collection permits, and wage labor sites enables spatial analysis of benefit flows. Zonal statistics can calculate average household income from JFM within specific buffer zones of forest patches or correlate benefit receipt with proximity to forest guards/FPC leaders, visually and statistically revealing spatial biases and leakages in the benefit-sharing mechanism (Behera and Engel 2006).

Land Tenure and Institutional Overlay Analysis: Digitizing layers of official JFM boundaries, Forest Rights Act (FRA) Community Forest Resource (CFR) claims, revenue land, and protected areas is crucial. GIS overlay analysis identifies spatial conflicts and tenure ambiguities areas where JFM overlaps with pending CFR claims or where FPC jurisdiction is disputed, directly impacting management legitimacy and community motivation (Springate-Baginski and Blaikie 2007).

Suitability Modeling for Intervention Planning: Multi-criteria Evaluation (MCE) within GIS integrates factors like forest degradation level, proximity to dependent villages, existing FPC strength, infrastructure access, and market linkages to model spatial priorities for JFM strengthening interventions or new site selection (Takam Tiamgne et al. 2022). In the Sundarbans, modeling incorporates salinity intrusion data and cyclone vulnerability to assess JFM resilience under climate change (Ghosh and Mistri 2023).

Spatially Explicit Monitoring Systems: Mobile GIS applications (e.g., ODK Collect, KoBoToolbox with GPS) enable FPC members or forest staff to geotag real-time data on patrol routes, encroachment incidents, plantation survival rates, or NTFP harvest locations. This creates dynamic spatial databases for adaptive management and transparent monitoring of JFM agreements (Longley et al. 2015).

Critically, GIS analysis must be contextually grounded. While powerful for revealing spatial patterns and quantifying biophysical changes, GIS alone cannot capture the full depth of power dynamics, cultural values, or subjective experiences within JFM. Integrating GIS findings with qualitative socio-political analysis is essential for robust interpretation. Furthermore, ethical considerations regarding data ownership, community consent for mapping sensitive resources or conflicts, and preventing the misuse of spatial data to restrict traditional access rights are paramount (Shit et al. 2021). When applied ethically and integrated with mixed methods, GIS becomes an indispensable toolkit for objectively assessing the spatially variegated landscape of JFM success and challenges in West Bengal, providing evidence crucial for equitable policy reform and adaptive management.

8.4 A Case Study on Joint Forest Management: Success Stories and Implementation Challenges in Arabari Forest Range

The Arabari Forest Range in Paschim Medinipur district of West Bengal stands as a pioneering example of participatory forest management in India, representing a transformative shift from exclusionary conservation policies to community-based resource governance (Fig. 8.4). Initiated in 1971–72 under the leadership of Forest Officer Ajit Kumar Banerjee, the Arabari experiment emerged as a response to severe forest degradation caused by decades of unrestricted extraction, inadequate protection mechanisms, and antagonistic relationships between forest officials and local communities (Poffenberger and McGean 1996). The forest compartment, spread across approximately 1272 hectares of degraded sal (*Shorea robusta*) forest, had been reduced to barren lands with sparse vegetation coverage below 20% by the early 1970s, primarily due to uncontrolled grazing, illegal felling, and lack of local stakeholder involvement in conservation efforts (Sarin 1993). The success of the Arabari model fundamentally rested on the innovative Forest Protection Committee (FPC) mechanism, which granted legal recognition to eleven village communities comprising predominantly tribal and marginalized populations, enabling them to protect and manage forest resources in exchange for usufruct rights, employment opportunities in silvicultural operations, and a significant share 25% of the final harvest proceeds (Saxena 1997). By the late 1980s, the forest crown density had increased remarkably to over 60%, biodiversity indices showed substantial improvement with regeneration of native species, and soil conservation was visibly enhanced, demonstrating the ecological viability of collaborative management approaches (Malhotra and Poffenberger 1989). However, implementation challenges have persisted throughout the program's evolution, including bureaucratic resistance to genuine power-sharing, inadequate financial allocation for FPC operations, exclusion of women from decision-making structures despite their significant dependence on non-timber forest products, elite capture of benefits by economically dominant community members, and conflicts over benefit-sharing mechanisms when commercial harvesting commenced (Ghate and Nagendra 2005). The limited tenurial security provided to communities typically through 5-year renewable agreements rather than permanent rights has undermined long-term conservation incentives, while the top-down microplanning processes have often marginalized indigenous ecological knowledge and traditional resource management practices (Sundar 2000). Furthermore, as Lele and Menon (2014) observe, the scaling up of the Arabari model across West Bengal through the 1990s led to significant dilution of its participatory essence, with many subsequent FPCs functioning merely as implementation agencies for forest department directives rather than autonomous decision-making bodies. The experience reveals that while Joint Forest Management can yield remarkable ecological restoration and livelihood improvements when implemented with genuine devolution of authority and equitable benefit distribution, its replication requires sustained institutional reforms, capacity building investments, gender-sensitive governance structures, and legal frameworks that recognize communities as legitimate custodians rather than temporary beneficiaries of forest resources (Bhattacharya et al. 2010).

The Arabari Forest Range in Paschim Medinipur district, West Bengal, represents the pioneering site of Joint Forest Management (JFM) in India, covering approximately 1272 hectares of sal (*Shorea robusta*) dominated moist deciduous forest. Located between 22°25’–22°35’N latitude and 87°10’–87°20’E longitude, the area features undulating terrain with lateritic soils and receives 1400–1600 mm annual rainfall (Malhotra and Poffenberger 1989; Bisui et al. 2023). Eleven villages surrounding the Arabari Forest Range house approximately 10,000–12,000 people, predominantly Santhal, Munda, and Bhumij tribal communities whose livelihoods depend critically on forest resources (Saxena 1997). By the early 1970s, intensive exploitation had reduced forest crown density below 20%, with unregulated grazing by 20,000 cattle and illegal felling transforming dense forests into degraded scrublands (Sundar 2000). The socio-economic context includes high poverty levels, with over 60% below-poverty-line households and average landholdings under 0.5 hectares (Bisui et al. 2023). Selected as an experimental JFM site in 1971–72 under Forest Officer A.K. Banerjee’s leadership, Arabari has become an internationally recognized model for community-based forest conservation and ecological restoration (Poffenberger and McGean 1996).

8.4.1 Database and Methodology

The empirical investigation of Joint Forest Management (JFM) dynamics and their socio-ecological ramifications necessitated a strategically designed sampling framework encompassing 140 rural households within the study area (Fig. 8.5). The research design incorporated a bifurcated sampling strategy, comprising equal proportions of households affiliated with JFM committees ($n = 70$) and those operating outside the formal JFM institutional structure ($n = 70$), systematically distributed across forest administrative jurisdictions. This deliberate methodological configuration facilitated a robust comparative assessment of divergent livelihood trajectories and ecosystem service access patterns, particularly within the historically significant Arabari Forest complex—recognized as a pioneering landscape in JFM implementation.

Fig. 8.4 Overview of the Arabari Forest Range: **a** The primary entryway to the Arabari Range, featuring a designated forest gate that marks the start of the protected area. **b** A green directional or boundary sign written, identifying the location as "Arabari." **c** Signposts indicating the presence of the Arabari Modern Nursery and the Research Range, which are central to the local forest conservation efforts. **d** The entrance to the administrative headquarters of the Forest Range Office, featuring informative murals or posters about local wildlife and nature conservation. **e** A dense view of the Sal trees (Shorea robusta) within the Arabari Range, showcasing the successful reforestation and regeneration of the canopy. **f** A road cutting through the heart of the forest, illustrating the lush greenery and the structured growth resulting from community-led forest management

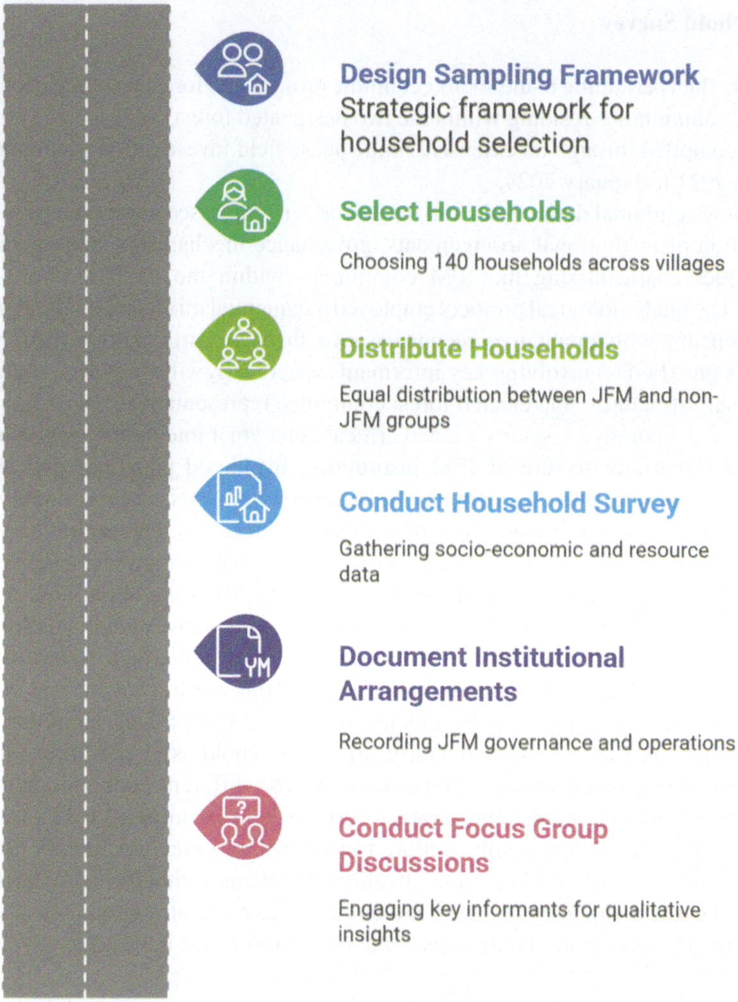

Fig. 8.5 Symmetrical diagram of the research methodology for the JFM study

The selected household units were spatially distributed across eight forest-fringe settlements: Ghagra, Kusumdhari, Soldiha, Salboni, Majighar, Chandmura, Burisol, and Ghuchisol. These villages represent characteristic socio-ecological interfaces where forest-dependent communities intersect with conservation governance frameworks. The dual-cohort sampling approach was specifically calibrated to illuminate the differential impacts of participatory forest governance mechanisms on resource access, livelihood security, and power asymmetries between institutionally integrated and excluded community segments. This methodological architecture thereby enables nuanced examination of how JFM operationalization translates into tangible outcomes across diverse stakeholder categories within the forest-society continuum.

Household Survey

Primary data pertaining to the socio-economic profiles and forest resource dependencies of communities residing within the two designated forest ranges were systematically compiled through an extensive multi-phase field investigation spanning from March 2021 to January 2024.

This longitudinal data acquisition framework encompassed comprehensive documentation of institutional arrangements, governance mechanisms, and operational modalities characterizing the JFM committees within the Arabari Forest landscape. The methodological protocol employed a sequential mixed-methods approach, commencing with qualitative reconnaissance through participatory focus group discussions (FGDs) involving key informants specifically village elders, traditional leadership structures, and elected forest committee representatives.

These deliberative sessions yielded critical contextual intelligence regarding the organizational architecture of JFM institutions, livelihood portfolios and income diversification strategies, residential infrastructure typologies, and community-level forest resource utilization patterns and management protocols. This preliminary qualitative phase established an empirically grounded foundation for understanding local governance dynamics and human-forest interactions. Building upon these foundational insights, structured household-level surveys were subsequently administered across settlements contiguous to the forest boundaries, utilizing a rigorously pretested and calibrated survey instrument. The questionnaire framework was conceptually designed to capture multi-dimensional data encompassing temporal livelihood transitions, income source stratification, household asset portfolios, and the extent of forest-based resource dependence across different economic activities. This integrated data architecture enables systematic assessment of JFM's transformative influence on community welfare trajectories, resource governance efficacy, and the sustainability of forest-based livelihood systems within the study landscape (Fig. 8.1). The comprehensive dataset thus generated facilitates granular analysis of participatory conservation outcomes at the household-forest interface.

Focus Group Discussions with Forest-Dependent Communities and Stakeholder Networks

To excavate nuanced understandings of Joint Forest Management's operational complexities and its multifaceted ramifications for forest-adjacent populations, participatory Focus Group Discussions (FGDs) constituted a pivotal methodological component of this investigation. These structured deliberative forums engaged diverse stakeholder constituencies, encompassing forest-dwelling households, communities inhabiting forest peripheries, and institutional actors embedded within the forest governance landscape (Fig. 5.3). The FGD framework was conceptually designed to elicit authentic perspectives, experiential narratives, and critical assessments regarding JFM's transformative effects from those positioned at the nexus of policy implementation and lived reality. The facilitation strategy deliberately cultivated an inclusive dialogic environment, encouraging uninhibited articulation of viewpoints and fostering collective knowledge construction around several thematic domains:

Livelihood Architecture and Forest Resource Dependence: Deliberations interrogated the intricate relationships between household economic portfolios and forest-derived resources, examining how participatory management regimes have reconfigured income generation pathways, resource accessibility patterns, and the broader contours of economic resilience and vulnerability within forest-dependent communities.

Governance Mechanisms and Resource Stewardship: Critical examination centered on JFM's operational effectiveness in achieving sustainable resource management outcomes, particularly scrutinizing the substantive inclusion of community actors in decision-making hierarchies, the ecological sustainability of extraction practices, and the alignment between traditional ecological knowledge and formal management protocols.

Power Asymmetries and Distributional Justice: A central analytical focus explored the micro-politics of forest governance, interrogating whether democratic principles pervade JFM committee functioning, assessing the inclusivity of marginalized social groups in deliberative processes, and evaluating the equity of benefit-sharing arrangements across differentiated community segments.

Constraints and Transformative Pathways: Participants critically identified systemic impediments undermining JFM efficacy including inter-community resource conflicts, institutional opacity, procedural barriers constraining meaningful participation, and governance deficits. Concurrently, discussions generated community-driven recommendations for institutional strengthening and adaptive management strategies.

Socio-ecological Outcomes and Community Transformation: Discourse extended to perceived environmental trajectories under JFM stewardship, encompassing biodiversity conservation effectiveness, forest regeneration dynamics, and ecosystem service provisioning. Social dimensions received parallel attention, examining shifts in customary resource-use practices, alterations to community solidarity structures, and the cultural implications of formalized forest governance.

The rich qualitative intelligence harvested through these participatory engagements provided essential interpretive depth, contextualizing and triangulating quantitative survey data. These insights illuminated critical intervention points for institutional refinement, enabling evidence-informed recommendations to enhance JFM's capacity to advance both ecological sustainability objectives and the livelihood security imperatives of forest-dependent communities.

Collection of Secondary Data

To establish a robust analytical foundation and contextually situate the primary empirical findings, systematic compilation of secondary data constituted an indispensable methodological dimension of this research endeavor. This archival component strategically augmented field-derived evidence, enabling comprehensive examination of Joint Forest Management's institutional evolution, policy architecture, and implementation trajectories within the study landscape. The secondary data acquisition strategy encompassed critical review and synthesis of authoritative documentation sourced from multiple institutional repositories. Official policy instruments, administrative records, and technical reports archived by the Ministry of Environment, Forest and Climate Change (MoEF & CC) formed a primary documentary corpus, furnishing insights into national-level forest governance frameworks, normative JFM operational guidelines, and evolving environmental conservation paradigms. Complementary documentation was systematically retrieved from the West Bengal Forest Department's administrative archives, including divisional working plans, forest range management records, JFM committee registration databases, and periodic progress reports detailing programmatic implementation at the state and district scales.

This documentary evidence yielded multi-dimensional information encompassing: the policy genealogy and regulatory architecture governing participatory forest management; institutional specifications defining JFM committee composition, membership criteria, and organizational hierarchies; longitudinal forest cover assessments and vegetation dynamics derived from Forest Survey of India inventories; and socio-demographic profiles characterizing forest-adjacent populations, including caste composition, occupational structures, and economic vulnerability indicators. Additional sectoral data pertaining to rural development interventions, watershed management programs, and natural resource-based livelihood initiatives were cross-referenced to comprehend the broader developmental context shaping forest-society interactions. This integration of secondary intelligence with primary field data enabled triangulation of findings, historical contextualization of contemporary governance arrangements, and critical assessment of policy-practice disjunctures characterizing JFM operationalization in the Arabari Forest complex.

Standard of Living Index (SLI)

To quantify household material welfare, a composite Standard of Living Index (SLI) was constructed using asset ownership and infrastructure access data collected through structured interviews. The analytical framework follows the National Family Health Survey (NFHS) methodology, incorporating 27 indicators across multiple dimensions: residential infrastructure quality, access to essential utilities (water, sanitation, cooking fuel, electricity), consumer durable ownership, and productive agricultural assets. Index computation employs the weighted aggregation methodology prescribed by the International Institute for Population Sciences (IIPS and ORC Macro 2000), where each component receives predetermined weights reflecting its contribution to overall welfare. This standardized approach generates a continuous scalar measure enabling inter-household comparisons and statistical assessment of JFM's impacts on living standards.

Table 8.1 Significance of forest resources for livelihood maintenance

Code	Description
D1	Can't do without
D2	Important
D3	Not so important
D4	Not important at all
D5	Don't know/no answer

Livelihood Dependency Index (LDI)

To quantify the role of forest products in sustaining household livelihoods, a Livelihood Dependency Index (LDI) was employed. This index classifies households according to their reliance on forest resources (Schmerbeck and Fiener 2015; Nerfa et al. 2020) (Table 8.1).

$$\begin{aligned} \text{LDI} =& (nD_1 \times 1) + (nD_2 \times 0.75) + (nD_3 \times 0.50) \\ &+ (nD_4 \times 0.25) + (nD_5 \times 0) \end{aligned} \tag{8.1}$$

where n represents the total number of surveyed households, whereas D_1, D_2, D_3, D_4, D_5, and D_n denote the various categories of forest products considered (Table 8.1).

Forest Dependency

Forest dependency was quantified by dividing total forest-derived income by overall household income (Zenteno et al. 2013).

$$\text{RFI} = \frac{\text{Total Forest Income (TFI)}}{\text{Total Household Income (THI)}} \tag{8.2}$$

where RFI represents the relative forest income, TFI denotes total income derived from forest resources, and THI indicates the overall household income.

Assessment of Ecosystem Services Valuation (ESV) and Ecosystem Functions (ESVf)

To systematically evaluate the economic magnitude and functional diversity of ecosystem services within the study landscape, forest cover was categorized into distinct density classes: dense forest and degraded forest formations, enabling differentiated valuation of ecological provisioning across varying vegetation conditions. Satellite imagery classification employed the maximum likelihood supervised classification algorithm, a robust parametric approach that probabilistically assigns pixels to predefined land cover categories based on spectral signature distributions, thereby ensuring accurate spatial delineation of target forest typologies.

The monetary valuation of ecosystem services and their constituent ecological functions was operationalized through adaptation of the seminal coefficient framework established by Costanza et al. (1997), which assigns standardized economic values to diverse ecosystem service categories including provisioning, regulating, supporting, and cultural services. The computational methodology integrated refinements proposed by Bisui et al. (2021) and Kindu et al. (2016), incorporating adjustments for regional ecological characteristics and socio-economic contexts. This integrated valuation framework facilitates quantitative assessment of forest ecosystems' multifunctional contributions to human welfare and environmental stability across the study area.

$$\mathrm{ESV} = \sum (A_k \times VC_k) \tag{8.3}$$

$$\mathrm{ESV}_f = \sum (A_k \times VC_{kf}) \tag{8.4}$$

where ESV denotes the total estimated value of ecosystem services, A_k represents the area (in hectares) of land-use/land-cover (LULC) type, VC_k refers to the value coefficient (US\$ ha^{-1} year^{-1}) associated with LULC type k, ESV_f indicates the ecosystem service function, and VC_{kf} represents the corresponding value coefficient of the ecosystem function for LULC type k.

Statistical Analysis

To rigorously examine Joint Forest Management's multi-dimensional impacts and operational effectiveness, a comprehensive suite of statistical analytical techniques was deployed. The analytical framework systematically evaluated JFM's influence across critical dimensions including gender inclusivity, forest product diversification, caste-based resource prioritization patterns, afforestation outcomes, vegetation cover dynamics, and community engagement intensity in participatory governance processes:

1. Descriptive Statistical Analysis: Summary statistics and frequency distributions were employed to synthesize stakeholder perceptions regarding diverse implementation aspects of JFM, providing foundational empirical characterization of program experiences and outcomes across the sample population.
2. F-Test for Variance Homogeneity: Two-sample F-tests were executed to assess whether statistically significant differences exist in the variance of sustainable forest management knowledge between JFM committee members and non-affiliated households, thereby testing the hypothesis that institutional participation influences knowledge acquisition and distribution.

3. Ordinal Scale Measurement: Participant satisfaction with JFM programmatic objectives was quantified using a 4-point Likert-type ordinal scale (Highly Satisfactory to Highly Unsatisfactory), enabling standardized measurement and comparative analysis of subjective program assessments across respondent categories.
4. Differential Participation Analysis: Comparative statistical examination explored variations in participatory engagement, decision-making authority, and institutional representation across socially stratified categories specifically gender, caste affiliation, and community identity illuminating systemic inequities and power asymmetries embedded within Forest Protection Committee structures and governance practices.

8.4.2 Result and Discussion

Access and Allocation of Forest Products in JFM Programs

The participatory framework of Joint Forest Management in the Arabari region demonstrates a nuanced pattern of forest resource utilization that reflects both subsistence necessities and income generation strategies among forest-fringe communities. The empirical evidence reveals a clear hierarchy of resource dependency, with firewood collection achieving universal participation (100%), underscoring its critical role as the primary energy source for rural households. This is complemented by remarkably high extraction rates for non-timber forest products (NTFPs), particularly fruits, leaves, and plants (93.75%) and rope-making materials (82.5%), indicating sophisticated traditional knowledge systems in resource processing and utilization.

The data illuminates a selective engagement pattern where communities prioritize readily accessible and renewable resources over more restricted products. Notably, timber and pole collection remain significantly limited (19% and 16.25%, respectively), suggesting effective regulatory mechanisms within the JFM framework that balance conservation imperatives with livelihood requirements (Table 8.2). The moderate participation in medicinal plant collection (70%) and honey harvesting (50%) reflects specialized knowledge distribution within the community, while the near-equal division in hunting activities (51.25%) and gum-resin collection (48.75%) indicate transitional resource-use patterns. The complete absence of grazing and farming activities within forest boundaries demonstrates successful boundary demarcation and institutional governance under Forest Protection Committees. The enhanced access to NTFPs, coupled with legal sanction for collection and commercialization, has created sustainable livelihood pathways while maintaining ecological integrity. This co-management model effectively synthesizes indigenous ecological wisdom with scientific forestry principles, resulting in both biodiversity conservation and socio-economic advancement of forest-dependent populations, thereby validating the JFM paradigm as a replicable framework for sustainable forest resource governance.

Table 8.2 Opinion of respondents regarding forest products collection in Arabari Forest Range

Forest resources	Yes (%)	No (%)
Collect timber	19	81
Collect poles	16.25	83.75
Make ropes	82.5	17.5
Collect firewood	100	0
Make charcoal	31.25	68.75
Collect medicines	70	30
Take fruits, leaves, plants	93.75	6.25
Collect honey	50	50
Collect gums and resin	48.75	51.25
Hunt for bush meat	51.25	48.75
Graze animals	0	0
Get wood for carving	33.75	66.25
Farming	0	100

Livelihood Dependency Index (LDI) Analysis in JFM Context

The Livelihood Dependency Index quantitatively illustrates the differentiated reliance patterns of forest-dependent communities on various forest products, revealing a stratified resource hierarchy that shapes socio-economic sustainability. Fuelwood emerges as the most critical resource with the highest LDI score of 73, reflecting its indispensable role in meeting daily energy requirements for cooking and heating in rural households lacking alternative energy infrastructure (Fig. 8.6). Timber follows with an LDI score of 59, indicating substantial dependence for construction and income generation purposes. Among non-timber forest products, sal leaves demonstrate remarkably high dependency (LDI 69), attributed to their commercial value in plate-making industries and traditional food packaging systems. Sal seeds (LDI 48) and Kendu leaves (LDI 46) represent moderate dependency levels, primarily serving as supplementary income sources during specific harvesting seasons. Medicinal plants (LDI 43) occupy an intermediate position, reflecting their dual role in traditional healthcare systems and market-oriented collection. The relatively lower dependency indices for honey (40), Mahua (37), and Mahua seeds (39) suggest their characterization as seasonal supplementary resources rather than primary livelihood anchors. This hierarchical dependency pattern underscores the multifunctional nature of forest ecosystems in supporting rural livelihoods while emphasizing the necessity for targeted management strategies that prioritize high-dependency resources within sustainable extraction frameworks under JFM governance structures.

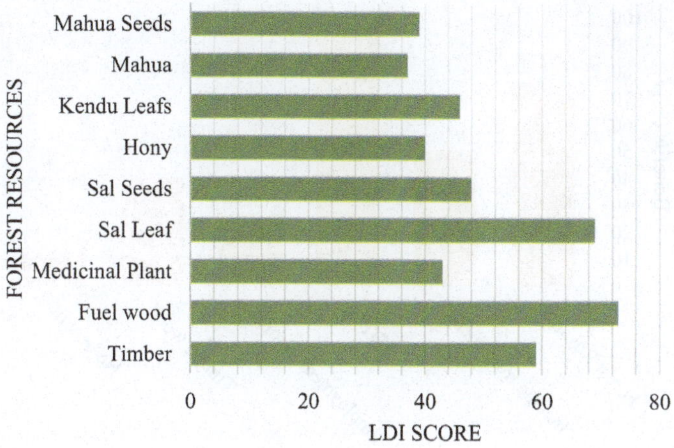

Fig. 8.6 Livelihood dependency on forest-based products

Benefit Distribution and Economic Outcomes of JFM Programs

The benefit distribution matrix from Joint Forest Management initiatives reveals a multi-dimensional livelihood enhancement framework wherein non-timber forest products constitute the predominant benefit stream, accessed by 94% of participating households. This overwhelming dependence on NTFPs underscores their critical importance as readily accessible, renewable resources that require minimal capital investment while providing year-round subsistence and market opportunities. Forest royalty benefits reach 76% of respondents, demonstrating effective revenue-sharing mechanisms between Forest Protection Committees and local communities, thereby ensuring equitable distribution of forest-generated income (Fig. 8.7). The moderate participation in social forestry nursery activities (51%) reflects specialized skill requirements and limited capacity absorption in plantation-related enterprises. Conversely, timber wood benefits remain constrained to 40% of households, indicative of regulated extraction protocols designed to balance commercial interests with conservation objectives. Employment generation in forest operations engages merely 27% of the population, suggesting limited absorption capacity in formal forestry activities and highlighting the predominance of self-employment through NTFP collection. Significantly, only 13% report receiving no benefits, affirming the inclusive nature of JFM frameworks. This benefit architecture demonstrates that successful participatory forest management transcends mere resource access, creating diversified income streams while maintaining ecological sustainability through controlled extraction and equitable allocation mechanisms that enhance rural prosperity.

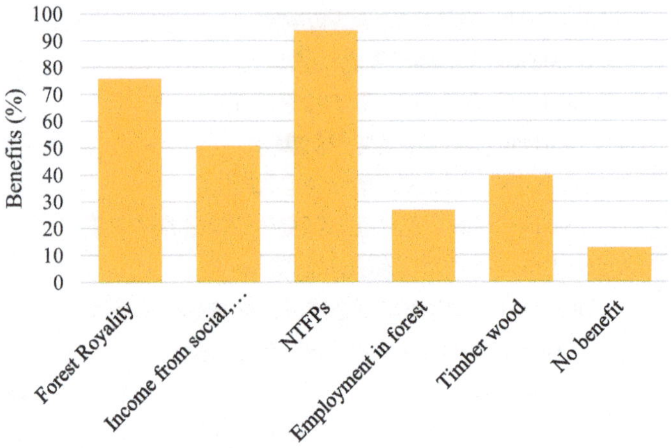

Fig. 8.7 Percentage benefit from forest (under JFM)

Standard of Living Index (SLI)

The Standard of Living Index was computed through weighted aggregation of household asset scores following the National Family Health Survey's standardized measurement protocol (IIPS and ORC Macro 2000). Empirical findings reveal a bimodal distribution of living standards across the surveyed settlements: approximately 68% of villages demonstrated moderate welfare levels (SLI scores ranging from 20 to 40), while 32% exhibited high living standards (SLI scores spanning 40–67). This distributional pattern suggests that forest resource-derived income streams constitute a substantial determinant of household material welfare within these rural communities, with forest-based economic activities contributing significantly to aggregate household income portfolios and consequently elevating overall living conditions among forest-dependent populations.

Community Perspectives on JFM's Role in Supporting Livelihoods and Forest Health

Community perceptions provide compelling evidence of Joint Forest Management's transformative impact on both ecological restoration and socio-economic advancement in the Arabari Forest landscape. An overwhelming majority (76.35%) perceive substantial improvement in forest conditions over the preceding 5-year period, with an additional 21.59% acknowledging moderate enhancement, collectively representing 97.94% positive assessment (Table 8.3). This near-universal recognition of ecological recovery validates the efficacy of participatory management approaches integrating local stewardship with scientific forestry practices. Critically, the direct correlation between forest health and livelihood improvement is evident, as 73.81%

Table 8.3 Respondent opinions on the effectiveness of joint forest management (JFM)

Structured questions	Type of responses	Responses in %
(a) How would you rate the change of the overall condition of Arabari Forest reserves today compared to 5 years ago?	Much improved	76.35
	Some improved	21.59
	Stable	2.06
(b) How has this change (improvement/worsening of forest condition) affected your livelihood?	It has much improved my livelihood	73.81
	It has much improved my livelihood a bit	17.36
	The change in forest condition did not have any effect on my livelihood	8.83
(c) Do you think rules restricting forest use are necessary to maintain the forest?	Yes	8346
	No	16.54

of respondents report significant livelihood enhancement attributable to improved forest conditions, while 17.36% note modest gains. The symbiotic relationship between conservation outcomes and economic welfare demonstrates that ecological restoration need not compromise rural prosperity but rather reinforces it through sustained resource availability and enhanced ecosystem services. Furthermore, the remarkable consensus supporting regulatory frameworks (83.46%) reflects evolved environmental consciousness and recognition that controlled access mechanisms ensure long-term resource sustainability rather than restricting opportunities. This institutional acceptance signifies successful internalization of conservation ethics among forest-dependent communities. The minimal proportion reporting no livelihood impact (8.83%) or opposing regulations (16.54%) represents inevitable heterogeneity in benefit distribution and risk perception, yet does not diminish the overwhelming validation of JFM as an effective paradigm for reconciling conservation imperatives with developmental aspirations in forest-fringe communities.

Assessing the Impact of JFM Implementation on Forest Ecosystem Services and Functions

The quantitative assessment of ecosystem service values in the Arabari Forest Ranges demonstrates the profound ecological and economic dividends of Joint Forest Management interventions over the three-decade period (1990–2023). The aggregate ecosystem service value increased from US$6.02 million to US$7.17 million, representing a net enhancement of US$1.15 million annually, thereby validating JFM's capacity to augment natural capital while supporting rural livelihoods (Table 8.4). This improvement is primarily attributed to the remarkable expansion of dense forest cover, whose ecosystem service value escalated from US$5.51 million to US$6.91 million a gain of US$1.4 million. Concurrently, degraded forest areas exhibited substantial contraction, with their associated ecosystem services declining from US$0.51 million to US$0.26 million, reflecting successful forest regeneration and conversion of degraded landscapes into productive ecosystems. The functional disaggregation reveals differential enhancement patterns across ecosystem services.

Climate regulation functions demonstrated the most substantial absolute gain (US$0.11 million), underscoring forests' critical role in carbon sequestration and microclimate stabilization. Erosion control services increased by US$0.10 million, highlighting improved soil conservation and watershed protection capabilities. Raw material provisioning enhanced by US$0.16 million, directly benefiting forest-dependent communities through sustained availability of timber and non-timber products. Water supply services remained stable, indicating consistent hydrological functions despite environmental pressures. Genetic resource values appreciated by US$0.02 million, reflecting biodiversity conservation achievements. However, food production services marginally declined by US$0.03 million, potentially indicating regulated access to prevent over-exploitation or shifting community dependencies toward alternative livelihood strategies (Table 8.5). The total ecosystem function value progression from US$2.74 million to US$3.10 million (gain of US$0.36 million) exemplifies how participatory forest governance synthesizes conservation objectives with socio-economic imperatives, creating resilient socio-ecological systems that deliver multiple benefits across temporal scales while maintaining ecological integrity for future generations.

Table 8.4 Estimated ecosystem service values (ESV in Million US$ per Year) for various years and changes (1990–2023) in Arabari Forest Ranges

Forest type	ESV (US$ Million)		ESV change (US$ Million)
	1990	2023	1990–2023
Dense forest	5.51	6.91	1.4
Degraded forest	0.51	0.26	-0.25
Total	6.02	7.17	1.15

Table 8.5 Annual estimated ecosystem function values (ESVf in million US$) and variations (1990–2023) for Arabari forest ranges

Ecosystem function	1990	2023	1990–2023
Food production	0.22	0.19	0.03
Raw materials	0.93	1.09	−0.16
Genetic resources	0.13	0.15	−0.02
Water supply	0.05	0.05	0
Climatic regulation	0.63	0.74	−0.11
Erosion control	0.78	0.88	−0.10
Total	2.74	3.10	0.36

Community Perceptions of Joint Forest Management: Evaluating Implementation and Power Dynamics

Community assessments of Joint Forest Management implementation across 20 Forest Protection Committee units reveal significant disparities between ecological achievements and socio-institutional equity outcomes. Afforestation initiatives receive overwhelming endorsement (76%), indicating successful plantation drives and community mobilization in forest restoration activities (Fig. 8.8). However, perceptions regarding forest cover increase (19%), forest product variety enhancement (17%), and plant species diversity (16%) remain disappointingly low, suggesting that quantitative afforestation efforts have not translated proportionately into qualitative improvements in forest composition or resource availability from the community perspective. This disconnect may reflect extended temporal lags between plantation establishment and productive maturity, or alternatively, monoculture plantation strategies that prioritize commercial species over biodiversity-rich ecosystems. Most critically, the data exposes profound governance deficiencies in social inclusivity dimensions. Gender equality within JFM structures garners merely 3% affirmative response, revealing entrenched patriarchal decision-making processes that systematically marginalize women's participation despite their substantial roles in forest resource collection and household management. Similarly, caste-based prioritization is negatively perceived by 98% of respondents, indicating either successful egalitarian implementation or persistent structural inequalities masked by nominal inclusivity. These findings underscore that while JFM demonstrates considerable success in ecological restoration, fundamental challenges persist in democratizing forest governance, ensuring equitable benefit distribution, and dismantling socio-cultural hierarchies that perpetuate exclusionary practices within participatory frameworks.

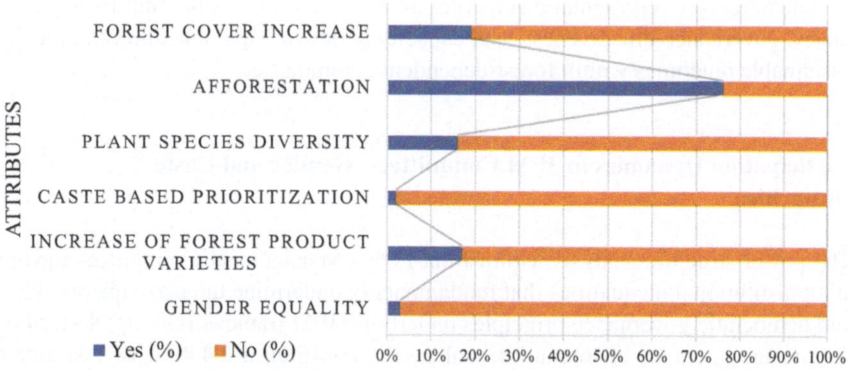

Fig. 8.8 People's response about the different attributes of JFM implication (n = 20 JFM FPC units)

Community Engagement in Joint Forest Management Across Forest Protection Committees (FPCs)

Empirical evidence reveals considerable heterogeneity in community participation patterns across distinct Joint Forest Management activities and organizational units. Attendance at formal FPC meetings and involvement in regulated timber harvesting operations demonstrate consistently robust participation rates across all committees examined, signifying widespread recognition of collective governance mechanisms and sustained commitment to collaborative forest stewardship. Similarly, household engagement in Non-timber Forest Product (NTFP) collection activities registers notably high levels, underscoring the critical subsistence and economic significance of these resources within local livelihood systems. However, participation intensity exhibits spatial and activity-specific variation, likely mediated by contextual factors including localized resource endowments, alternative economic opportunities, socio-cultural norms governing resource access, and historical patterns of forest dependency.

Gender dimensions of participatory governance present a nuanced landscape. While female participation in deliberative forums registers as substantive across several FPCs with certain committees demonstrating particularly elevated women's attendance rates inter-committee disparities persist, revealing incomplete realization of gender-inclusive governance ideals. These differential engagement patterns suggest that structural barriers, cultural constraints, or institutional design features continue to circumscribe women's equitable integration into decision-making spaces across the JFM institutional landscape. The household-level data collectively illuminates the active mobilization of FPC constituencies in diverse forest management interventions while simultaneously exposing critical imperatives for strengthening participatory mechanisms. Enhancing consistent, meaningful, and socially inclusive engagement particularly advancing women's substantive participation beyond tokenistic representation emerges as essential for consolidating JFM's legitimacy, governance effectiveness, and capacity to deliver equitable and ecologically sustainable outcomes within forest-dependent communities.

Participation Dynamics in JFM Committees: Gender and Caste Disparities

The power structure analysis within Joint Forest Management Committees unveils stark institutional inequalities that fundamentally undermine the participatory ethos and democratic governance principles underlying JFM frameworks (Fig. 8.9). Male dominance pervades all organizational levels, constituting 94.49% of committee membership, with their control intensifying across decision-making hierarchies from 91.48% meeting attendance to 69.84% representation in decision-making bodies. This hegemonic masculine participation contrasts sharply with abysmal female engagement, where women comprise merely 5.51% of membership, declining precipitously to 3.31% in meeting attendance and catastrophically to 1.76% in both

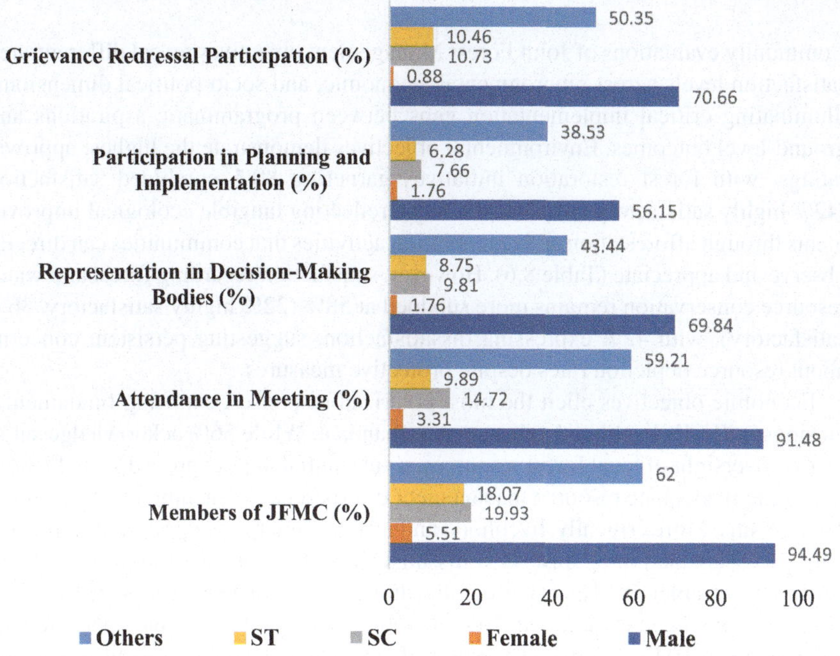

Fig. 8.9 Power structure and imbalance analysis across attributes, categories, and divisions of JFMc in the forest division

decision-making representation and planning implementation roles. Most alarmingly, female participation in grievance redressal mechanisms stands at a negligible 0.88%, effectively silencing women's voices in conflict resolution despite their substantial dependence on forest resources for household sustenance.

Caste-based stratification further compounds these exclusionary patterns. Scheduled Castes and Scheduled Tribes, historically marginalized communities with profound forest dependencies, represent only 19.93% and 18.07% of membership, respectively, with their participation eroding systematically across engagement levels SC participation diminishes from 14.72% meeting attendance to 7.66% in planning activities, while ST engagement contracts from 9.89% to 6.28% across comparable metrics. The "Others" category, presumably representing dominant castes, monopolizes institutional spaces with 62% membership and disproportionate control over decision-making (43.44%) and grievance mechanisms (50.35%). This intersectional analysis reveals that JFM committees, rather than democratizing forest governance, have institutionalized pre-existing social hierarchies, creating exclusionary spaces where elite male members from dominant castes exercise hegemonic control over forest resources and benefits, thereby perpetuating historical marginalization patterns and contradicting the inclusive, equitable participation principles that theoretically underpin community-based natural resource management paradigms.

Assessment of Respondent Opinions on Forest Management Objectives

Community evaluations of Joint Forest Management objectives reveal differentiated satisfaction levels across environmental, economic, and socio-political dimensions, illuminating critical implementation gaps between programmatic aspirations and ground-level outcomes. Environmental objectives demonstrate the highest approval ratings, with forest restoration initiatives garnering 80% combined satisfaction (42% highly satisfactory, 38% satisfactory), reflecting tangible ecological improvements through afforestation and regeneration activities that communities can directly observe and appreciate (Table 8.6). However, satisfaction regarding forest and water resource conservation remains more subdued at 58% (22% highly satisfactory, 36% satisfactory), with 42% expressing dissatisfaction, suggesting persistent concerns about resource depletion rates despite protective measures.

Economic objectives elicit the most polarized responses, exposing fundamental tensions in JFM's livelihood enhancement mandate. While 56% acknowledge satisfactory oversight of local forest resources, a substantial 44% express dissatisfaction, indicating inadequate resource management transparency or inequitable distribution mechanisms. More critically, livelihood and income generation opportunities receive only 51% satisfaction against 49% dissatisfaction the narrowest approval margin across all parameters. This near-equilibrium reflects uneven benefit distribution, limited employment creation, and possibly elite capture of economic opportunities, contradicting JFM's poverty alleviation objectives. The socio-political dimension reveals the most profound governance deficit, with merely 46% satisfaction regarding community empowerment and usufruct rights compared to 54% dissatisfaction (36% unsatisfactory, 18% highly unsatisfactory). This majority dissatisfaction underscores persistent power asymmetries where local communities perceive decision-making authority remaining concentrated within forest bureaucracy despite nominal devolution frameworks. The 18% highly unsatisfactory rating the highest negative intensity across all objectives signals acute frustration with tokenistic participation structures that grant procedural involvement without substantive decisional autonomy. These findings collectively suggest that while JFM achieves measurable ecological restoration, it substantially underperforms in democratizing forest governance and ensuring equitable economic benefits, necessitating fundamental institutional reforms to genuinely empower marginalized forest-dependent communities.

Knowledge Disparities in Sustainable Forest Management Between JFMC and Non-JFMC Members

The comparative knowledge assessment between Joint Forest Management Committee members and non-members reveals profound capacity building differentials attributable to institutional engagement and targeted training interventions within participatory forest governance frameworks. JFMC members demonstrate substantially superior technical competencies across multiple sustainable forest

Table 8.6 Perception of respondents about JFM Program implementation ($n = 140$) using Likert 4-point scale

Objectives	Highly satisfactory opinions (%)	Satisfactory opinions (%)	Unsatisfactory opinions (%)	Highly unsatisfactory opinions (%)
Environmental				
To safeguard and sustain the existing but rapidly depleting forests and water resources	22	36	30	12
To promote the restoration of degraded forest areas in order to expand green cover	42	38	12	8
Economic				
To effectively oversee and utilize local forest and water resources	24	32	34	10
To provide the forest-dependent communities with opportunities for livelihood and income generation	27	24	31	18
Socio-political				
Empower local communities to make decisions about forest use by granting them usufruct rights to forest lands	12	34	36	18

management dimensions, with the most dramatic disparities observed in specialized forestry practices. Plantation management knowledge exhibits the widest gap, with 95.43% of JFMC members possessing competency compared to merely 38.79% of non-members a differential of 56.64 percentage points indicating intensive skill development through active participation in afforestation programs (Table 8.7).

Similarly, fire management expertise shows remarkable divergence (75.85% versus 15.25%), reflecting systematic training protocols that equip committee members with critical protective capabilities essential for forest conservation. Exclusive knowledge domains emerge in sustainable non-timber forest product management, where 8.65% of JFMC members possess understanding while non-members demonstrate complete absence of such specialized knowledge, highlighting niche

Table 8.7 Knowledge of JFMC and Non-JFMC members about sustainable forest management ($N = 140$)

Attributes	JFMC members (%)	Non-JFMC members (%)
Environmental awareness programmed	82.12	78.63
Plantation management	95.43	38.79
Fire management	75.85	15.25
Biodiversity conservation	38.34	32.54
Soil and moisture conservation	19.47	13.84
Deforestation and land	98.05	97.12
Conservation of NTFPs	18.74	8.06
Sustainable non-timber forest	8.65	0

technical capacity development within institutional frameworks. Environmental awareness programs show relatively minimal disparity (82.12% versus 78.63%), suggesting broader community-level dissemination efforts that transcend formal membership boundaries. Interestingly, deforestation awareness demonstrates near-universal comprehension across both cohorts (98.05% and 97.12%), indicating effective general environmental education initiatives. However, critical knowledge gaps persist even among JFMC members in biodiversity conservation (38.34%), soil and moisture conservation (19.47%), and NTFP conservation (18.74%), revealing inadequate training curricula that prioritize production-oriented forestry over ecological sustainability principles. The F-test analysis ($F = 1.18$, $p = 0.41$) demonstrates statistically non-significant variance between groups despite substantial mean differences (142.8 vs. 90.375), suggesting heterogeneous knowledge distribution within both populations (Table 8.8). These findings underscore that while JFM membership facilitates enhanced technical capacity through experiential learning and institutional support, comprehensive ecological literacy remains deficient, necessitating holistic training frameworks that integrate biodiversity conservation, soil–water management, and ecosystem service principles alongside traditional silvicultural practices to achieve genuine sustainable forest management outcomes.

Table 8.8 F-test among knowledge of JFMC and non-JFMC members

Tests	JFMc members (%)	Non JFMc members (%)
Mean	142.8	90.375
Variance	9626.1	8141.6
Observations	8	8
df	7	7
F	1.18	
P(F < = f) one-tail	0.41***	
F Critical one-tail	3.787	

8.4.3 Conclusion

The Arabari Joint Forest Management experiment represents a seminal yet paradoxical achievement in participatory natural resource governance, simultaneously embodying transformative possibilities and persistent structural contradictions. The empirical evidence unequivocally demonstrates JFM's remarkable ecological restoration capacity, with forest density escalating from critically degraded conditions (<20% canopy cover) in the early 1970s to over 60% by the late 1980s, while ecosystem service valuations increased by US$1.15 million annually between 1990 and 2023. Equally significant are tangible livelihood enhancements, with 94% of households accessing non-timber forest products and 76% receiving forest royalty benefits, affirming that conservation and rural prosperity need not constitute zero-sum imperatives when participatory frameworks genuinely integrate community needs. However, beneath these celebratory narratives lie profound governance deficits that fundamentally compromise JFM's emancipatory claims. The power structure analysis reveals systematic exclusion of marginalized constituencies, with women comprising merely 5.51% of committee membership, declining catastrophically to 0.88% in grievance redressal mechanisms, while Scheduled Castes and Scheduled Tribes represent only 19.93% and 18.07% of membership, respectively. This institutionalization of patriarchal and caste-based hierarchies creates extractive governance spaces where dominant-caste male elites monopolize benefits and decisional authority. Community satisfaction assessments expose critical implementation gaps: while environmental restoration garners 80% approval, socio-political empowerment objectives receive majority dissatisfaction (54%), with merely 46% perceiving genuine devolution of decision-making autonomy. JFM's sustainability ultimately hinges on fundamental institutional reforms addressing power asymmetries, ensuring substantive participation of women and marginalized castes, guaranteeing transparent benefit distribution, and genuinely devolving authority from forest bureaucracies to community institutions. The Arabari legacy underscores that participatory conservation's legitimacy rests not on ecological metrics alone, but on its capacity to democratize environmental governance and create equitable socio-ecological futures where forest-dependent communities emerge as rights-bearing custodians rather than regulated beneficiaries of their ancestral landscapes.

References

Agarwal B (2009) Gender and forest conservation: the impact of women's participation in community forest governance. Ecol Econ 68(11):2785–2799

Agarwal B (2019) The interplay of ideas, institutional innovations and organisational structures: insights from group farming in India

Agrawal A, Ostrom E (2001) Collective action, property rights, and decentralization in resource use in India and Nepal. Polit Soc 29(4):485–514

Agrawal A, Gupta K (2005) Decentralization and participation: the governance of common pool resources in Nepal's Terai. World Dev 33(7):1101–1114

Agrawal A, Chhatre A (2006) Explaining success on the commons: community forest governance in the Indian Himalaya. World Dev 34(1):149–166

Behera B, Engel S (2006) Institutional analysis of evolution of joint forest management in India: a new institutional economics approach. Forest Policy Econ 8(4):350–362

Bhattacharya P, Pradhan L, Yadav G (2010) Joint forest management in India: experiences of two decades. Resour Conserv Recycl 54(8):469–480

Bisui S, Roy S, Bera B, Adhikary PP, Sengupta D, Bhunia GS, Shit PK (2023) Economical and ecological realization of Joint Forest Management (JFM) for sustainable rural livelihood: a case study. Trop Ecol 64(2):296–306

Bisui S, Roy S, Sengupta D, Bhunia GS, Shit PK (2021) Assessment of ecosystem services values in response to land use/land cover change in tropical forest. In: Forest resources resilience and conflicts. Elsevier, pp 435–447

Biswas T (2010) A spatio-temporal analysis of landscape change within the Eastern Terai, India: linking grassland and forest loss to changes in river course and land use. Utah State University.

Costanza R, d'Arge R, De Groot R, Farber S, Grasso M, Hannon B, Van Den Belt M (1997) The value of the world's ecosystem services and natural capital. Nature, 387(6630):253–260

DasGupta R, Shaw R (2017) Perceptive insight into incentive design and sustainability of participatory mangrove management: a case study from the Indian Sundarbans. J Forestry Res 28(4):815–829

Edmunds DS, Wollenberg EK (eds) (2013) Local forest management: the impacts of devolution policies. Routledge

IIPS and ORC Macro (2000) National Family Health Survey (NFHS-2), 1998–99: India. International Institute of Population Sciences, Mumbai

Ghate R, Nagendra H (2005) Role of monitoring in institutional performance: forest management in Maharashtra, India. Conserv Soc 3(2):509–532

Ghate R (2003) Ensuring 'collective action'in 'participatory'forest management. The Institute of Development Studies and Partner Organisations

Ghosh S, Mistri B (2023) Cyclone-induced coastal vulnerability, livelihood challenges and mitigation measures of Matla-Bidya inter-estuarine area, Indian Sundarban. Natural Hazards 116(3):3857–3878

Kindu M, Schneider T, Teketay D, Knoke T (2016) Changes of ecosystem service values in response to land use/land cover dynamics in Munessa-Shashemene landscape of the Ethiopian highlands. Sci Total Environ 547:137–147

Larson AM (ed) (2010) Forests for people: Community rights and forest tenure reform. Earthscan

Lele SM, Menon A (eds) (2014) Democratizing forest governance in India (p. 432). Oxford University Press, New Delhi

Lele SR, Nadeem K, Schmuland B (2010) Estimability and likelihood inference for generalized linear mixed models using data cloning. J Am Stat Assoc 105(492):1617–1625

Longley PA, Goodchild MF, Maguire DJ, Rhind DW (2015) Geographic information science and systems. John Wiley & Sons

Malhotra KC, Poffenberger M (1989) Forest regeneration through community protection: the West Bengal experience

Malhotra KC, Debal Deb DD, Dutta M, Vasulu TS, Yadav G, Adhikari M (1993) The role of non-timber forest products in village economies of south-west Bengal

Murali KS, Murthy IK, Nagaraja BC, Ravindranath NH (2003) A decade of JFM in India and its ecological impacts in. Root to Canopy–Regenerating forests through community–state partnerships, 275–296

Nerfa L, Rhemtulla JM, Zerriffi H (2020) Forest dependence is more than forest income: development of a new index of forest product collection and livelihood resources. World Dev 125:104689

Poffenberger M, McGean B (1996) Village voices, forest choices. Joint forest management in India Oxford University Press Delhi

Sarin M (1993) From conflict to collaboration: local institutions in joint forest management (No. 14). National Support Group for Joint Forest Management, Society for Promotion of Wastelands Development & the Ford Foundation

Saxena NC (1997) The saga of participatory forest management in India. CIFOR

Schmerbeck J, Fiener P (2015) Wildfires, ecosystem services, and biodiversity in tropical dry forest in India. Environ Manage 56(2):355–372

Shit PK, Pourghasemi HR, Adhikary PP, Bhunia GS, Sati VP (eds) (2021). Forest resources resilience and conflicts. Elsevier

Soubry I, Doan T, Chu T, Guo X (2021) A systematic review on the integration of remote sensing and GIS to forest and grassland ecosystem health attributes, indicators, and measures. Remote Sensing 13(16):3262

Springate-Baginski O, Blaikie PIERS (2007) Forests. People and power: the political ecology of reform in South Asia, (ed.), Earthscan, London

Sundar N (2000) Unpacking the 'joint'in joint forest management. Dev Chang 31(1):255–279

Takam Tiamgne X, Kanungwe Kalaba F, Raphael Nyirenda V, Phiri D (2022) Modelling areas for sustainable forest management in a mining and human dominated landscape: a geographical information system (GIS)-multi-criteria decision analysis (MCDA) approach. Ann GIS 28(3):343–357

Watson RT, Noble IR, Bolin B, Ravindranath NH, Verardo DJ, Dokken DJ (2000) Land use, land use change, and forestry, pp ix+-373

Zenteno M, Zuidema PA, de Jong W, Boot RG (2013) Livelihood strategies and forest dependence: new insights from Bolivian forest communities. Forest Policy Econ 26:12–21

Chapter 9
Climate Change Impacts on Forest Landscapes and Community Adaptations

Abstract The Sundarbans mangrove ecosystem has experienced unprecedented climate-driven transformation between 1990 and 2024, serving as a critical sentinel for understanding deltaic vulnerability under accelerating environmental change. This comprehensive analysis synthesizes multi-temporal satellite imagery, ground-based monitoring, and socio-economic data to document climate change impacts on the world's largest contiguous mangrove forest. Regional sea-level rise (3.2–8.0 mm/year), substantially exceeding global averages, has driven land loss exceeding 170 km^2 with complete submergence of islands including Ghoramara and Lohachara. Intensified tropical cyclones generated storm surges exceeding 5 m, causing catastrophic saline inundation. Surface water salinity increased by 15–20% between 1990 and 2020, devastating the keystone species *Heritiera fomes* (Sundari), which declined by > 70% in core habitats where salinity tolerance thresholds were exceeded. Rising temperatures (0.8–1.2 °C increase), altered precipitation patterns, and ocean warming (0.7–1.0 °C SST increase) have contributed to ecosystem stress and declining fishery resources. These synergistic climate stressors triggered fundamental reorganization from diverse, structurally complex Sundari-dominated forests to simplified stands of salt-tolerant pioneer species. Remote sensing analysis reveals net loss of 124 km^2 (5.2%) mangrove forest cover, with total forest-to-non-forest conversion reaching 1084.29 km^2 between 1990 and 2024. Spatial heterogeneity reflects differential vulnerability, with western areas maintaining better persistence due to freshwater inputs while central and eastern zones experienced severe degradation. Community responses evolved from reactive strategies to sophisticated adaptation measures, yet climate change acceleration consistently outpaced adaptation efforts, driving agricultural collapse and climate-induced displacement. The documented transformation represents threshold-crossing behavior where multiple stressors interact synergistically, producing non-linear ecosystem responses and positive feedback loops amplifying vulnerability. This study demonstrates that traditional conservation approaches are insufficient under rapid climate change, requiring transformation-oriented management strategies. The Sundarbans case provides critical insights for global mangrove conservation, illustrating cascading implications for carbon cycling, coastal protection, and biodiversity conservation under unprecedented environmental change.

P. K. Shit and M. Hasanuzzaman, *Forest Landscape Dynamics in West Bengal*,
Advances in Asian Human-Environmental Research,
https://doi.org/10.1007/978-3-032-23700-2_9

Keywords Sundarbans · Mangrove ecosystem · Climate change impacts · Sea-level rise · Tropical cyclones · Salinity intrusion · *Heritiera fomes* · Ecosystem transformation · Adaptation strategies · Deltaic vulnerability

9.1 Introduction

The accelerating pace of anthropogenic climate change is fundamentally altering forest ecosystems globally, imposing multifaceted biophysical stresses that cascade into significant socio-economic challenges for dependent communities (Malik et al. 2024). Rising mean temperatures and altered precipitation patterns are driving increased moisture stress, particularly in semi-arid and Mediterranean climates, leading to widespread tree mortality events, reduced growth rates, and shifts in species composition and distribution ranges (Allen et al. 2010; IPCC 2022). Concurrently, the hydrological cycle intensification manifests in more frequent and severe droughts in some regions, exacerbating water stress for vegetation, while other areas face increased flooding and soil saturation, disrupting root function and nutrient cycling. These primary stressors amplify the frequency, intensity, and spatial extent of natural disturbances. Climate change acts as a key driver in the unprecedented scale of wildfires, fueled by drier fuels and longer fire seasons, particularly in boreal and western North American forests (Abatzoglou and Williams 2016; Seidl et al. 2017). Similarly, warmer temperatures facilitate the expansion and proliferation of native and invasive insect pests (e.g., bark beetles) and pathogens, leading to extensive forest dieback and increased vulnerability to subsequent disturbances (Raffa et al. 2008). The synergistic effects of these impacts drought, heat stress, wildfires, and pests—are driving biome shifts, potential forest dieback in vulnerable areas like parts of the Amazon, and a reduction in overall forest resilience, fundamentally changing the structure, function, and services these landscapes provide (IPCC 2022).

These profound ecological transformations have direct and often severe consequences for human communities whose livelihoods, cultural identity, and well-being are intrinsically linked to forest ecosystems (Kofinas and Chapin 2009). Resource-dependent communities, particularly Indigenous peoples and local populations, face heightened vulnerabilities as climate change disrupts the availability and predictability of critical forest resources like timber, non-timber forest products (NTFPs), game, and medicinal plants (Ford 2012). This impacts subsistence economies, traditional food systems, and culturally significant practices. Increased wildfire smoke poses significant public health risks, while flooding and landslides linked to deforestation and extreme rainfall events threaten settlements and infrastructure (McDowell et al. 2020). Furthermore, climate-driven shifts can undermine culturally significant landscapes and species, eroding traditional ecological knowledge systems and spiritual connections to the land (Whyte 2017). The capacity of communities to adapt is highly uneven, often constrained by factors like poverty, limited access to alternative livelihoods, inadequate infrastructure, land tenure insecurity, and insufficient institutional support, highlighting issues of climate justice (Adger et al. 2013).

In response, a diverse array of adaptation strategies is emerging, spanning ecological management and socio-institutional approaches. Forest management increasingly focuses on enhancing ecosystem resilience through practices like thinning to reduce fuel loads and drought stress, promoting structural and species diversity, implementing assisted migration of climate-adapted genotypes, and restoring natural fire regimes where feasible (Millar et al. 2007; Nagel et al. 2017). Community-based adaptation (CBA) is crucial, emphasizing local agency and knowledge. Strategies include diversifying livelihoods away from sole forest dependence (e.g., agroforestry, ecotourism), revitalizing or adapting traditional fire management practices (e.g., cultural burning), strengthening early warning systems for fires and pests, developing community forests for enhanced local stewardship, and establishing networks for sharing adaptation knowledge (Chhatre and Agrawal 2009; Fernandez-Llamazares et al. 2021). Effective adaptation necessitates robust governance frameworks that integrate scientific knowledge with local and Indigenous knowledge systems (IK), ensure equitable access to resources and decision-making, provide adequate financial and technical support, and foster collaborative landscape-level planning that bridges ecological and social resilience goals (Berkes 2017; IPCC 2022). The success of these adaptations will determine the future sustainability of both forest ecosystems and the human communities intertwined with them.

9.2 Climate Change Impacts on Forest Landscapes and Community Adaptations in West Bengal

West Bengal's diverse forest ecosystems spanning the vulnerable mangrove forests of the Sundarbans, the biodiverse Terai-Duars foothills, the Sal-dominated tracts of the Chota Nagpur Plateau extension, and the fragile Himalayan forests in Darjeeling and Kalimpong face intensifying pressures from anthropogenic climate change, manifesting in regionally specific impacts and driving complex community adaptation responses (Das 2021) (Fig. 9.1). In the Sundarbans, a UNESCO World Heritage Site and the world's largest contiguous mangrove forest, rising sea levels (projected at 3–8 mm/year regionally) coupled with increased salinity intrusion due to reduced freshwater flow from upstream and more intense cyclonic events (e.g., Mondal et al. 2021) are causing significant mangrove dieback, particularly of the salinity-sensitive *Heritiera fomes* (Sundari), and altering species composition toward more salt-tolerant varieties (Hazra et al. 2002; IPCC 2022). This ecological degradation directly undermines the critical storm surge buffering capacity of the mangroves and depletes vital fishery and honey resources, severely impacting the livelihoods and safety of millions residing in the delta. Concurrently, the Himalayan and sub-Himalayan forests experience altered precipitation patterns, with observed increases in erratic, high-intensity rainfall events triggering landslides and soil erosion, while warmer temperatures facilitate the upward migration of pests and pathogens and alter phenological cycles of economically important species like cardamom and tea (Ravindranath et al. 2011; Tambe et al. 2012). In the drier, deciduous forests of Purulia, Bankura, and West Midnapore districts, rising temperatures and increased frequency of droughts are exacerbating water stress, reducing forest productivity, increasing vulnerability to fires, and impacting Non-Timber Forest Product (NTFP) availability crucial for tribal communities like the Santhals and Mundas (Basu 2020).

These climate-driven ecological shifts impose severe socio-economic burdens on forest-dependent communities across the state. Sundarbans communities face heightened risks of displacement, loss of fishing grounds, salinization of agricultural land, and increased human-wildlife conflict as stressed tigers range closer to settlements (Das 2021). In the Himalayan region, landslides damage infrastructure and agriculture, while changes in forest productivity threaten traditional livelihoods based on NTFPs and agroforestry. Plateau communities experience reduced availability of key NTFPs (e.g., Sal leaves, Kendu leaves, Mahua flowers, lac) essential for income and nutrition, compounded by water scarcity and crop failures (Kumar and Saikia 2020). Women, often primary NTFP collectors, and landless laborers are disproportionately affected. Furthermore, climate impacts erode cultural practices

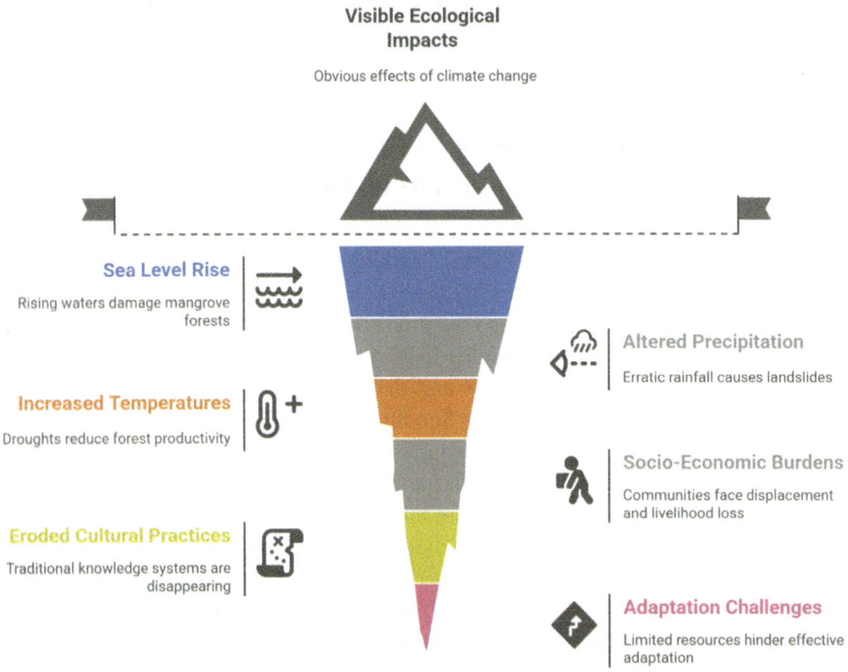

Fig. 9.1 Symmetrical diagram of climate change impacts on West Bengal's Forest ecosystems and communities

and traditional ecological knowledge systems intrinsically linked to specific forest species and seasonal cycles (Basu 2020).

Adaptation strategies are emerging, often blending traditional knowledge with external interventions, though significant challenges remain. In the Sundarbans, communities are adapting through constructing and raising homestead plinths, transitioning to saline-tolerant rice varieties and aquaculture (e.g., crab fattening, brackish water shrimp), diversifying livelihoods (e.g., small trade, migration), restoring mangrove belts, and participating in early warning systems (DasGupta and Shaw 2013). However, hard embankments often prove maladaptive, increasing vulnerability during breaches. Himalayan communities are reviving traditional water management (spring rejuvenation, ponds), adopting soil conservation techniques, diversifying crops, and exploring ecotourism. Plateau communities focus on water conservation (check dams, rainwater harvesting), NTFP value addition, exploring drought-resistant crops, and community forest management under initiatives like the Joint Forest Management (JFM) program (Tiwary 2019). State-led interventions include mangrove afforestation, cyclone shelters, JFM strengthening, and promoting alternative livelihoods. However, effective adaptation is often hampered by limited financial resources, inadequate technical capacity at the local level, institutional fragmentation, land tenure insecurity, insufficient integration of Indigenous knowledge into formal planning, and the sheer scale of the climate challenge, particularly for the most marginalized communities. Building true resilience necessitates enhanced cross-sectoral governance, equitable access to resources and decision-making, significant investment in locally appropriate adaptation infrastructure and capacity building, and deeper integration of scientific and traditional knowledge systems.

9.3 Integrating Geospatial Tools: Quantifying Climate Impacts and Informing Forest Community Adaptations

Geographic Information Systems (GIS) and associated geospatial technologies provide indispensable frameworks for quantifying the multifaceted impacts of climate change on forest landscapes and evaluating the efficacy of community adaptation strategies (Table 9.1). By integrating diverse spatial datasets including time-series satellite imagery (e.g., Landsat, Sentinel), climate model projections (downscaled CMIP6 data), digital elevation models (DEMs), soil maps, ecological inventories, and socio-economic data GIS enables the analysis of complex spatial patterns and processes driving forest vulnerability. Key quantification techniques include:

Vegetation Change Detection: Utilizing spectral indices like NDVI (Normalized Difference Vegetation Index), EVI (Enhanced Vegetation Index), and NBR (Normalized Burn Ratio) derived from satellite data to map and monitor long-term trends in forest health, productivity, phenology shifts, drought stress (e.g., using NDMI-Normalized Difference Moisture Index), and post-disturbance recovery (e.g., after

Table 9.1 Key GIS techniques for quantifying climate impacts on forests

Impact type	Primary GIS/geospatial techniques	Key outputs/indicators
Vegetation health and change	Time-series spectral indices (NDVI, EVI, NDMI, NBR), machine learning classification (land cover change), Lidar structural analysis	Biomass trends, canopy moisture stress, degradation extent, species shift maps, disturbance history
Climate exposure	Spatial interpolation, Downscaling of CMIP projections, drought index mapping (SPI/SPEI)	Temperature/precipitation anomaly maps, drought severity zones, future climate envelopes
Disturbance mapping	dNBR (fire severity), Lidar change detection, pest/host suitability modeling, cyclone inundation modeling	Burn severity classes, windthrow damage maps, outbreak risk maps, storm surge extent
Vulnerability assessment	Multi-criteria decision analysis (MCDA), spatial overlay of exposure/sensitivity/adaptive capacity layers	Composite vulnerability indices, risk hotspots for ecosystem

cyclones or fires). Machine learning algorithms (e.g., Random Forest, Support Vector Machines) applied within GIS platforms classify land cover changes, detect forest degradation, and identify areas of significant biomass loss or species composition shifts over decades (Goetz et al. 2010; Zhu 2017).

Climate Exposure Mapping: Spatial interpolation of historical climate station data and downscaling of future climate projections (temperature, precipitation, drought indices like SPI/SPEI, extreme event frequency) onto forest landscapes. GIS overlays identify forests experiencing anomalously high rates of warming, declining rainfall, or increased drought severity, highlighting areas of high *exposure* (Wang et al. 2015).

Vulnerability and Risk Modeling: Developing spatially explicit vulnerability indices by combining layers representing *exposure* (climate stressors), *sensitivity* (e.g., forest type, slope, soil erodibility, proximity to edge) and *adaptive capacity* (e.g., community socio-economic status, infrastructure access, land tenure security, institutional support). Multi-Criteria Decision Analysis (MCDA) within GIS is frequently employed to weight and integrate these diverse factors, mapping hotspots of high climate risk for specific forest values (biodiversity, carbon, NTFPs) and dependent communities (O'Brien et al. 2004; Metcalf et al. 2015).

Disturbance Impact Assessment: GIS is crucial for mapping the extent and severity of climate-amplified disturbances. Burn severity analysis using dNBR (differenced NBR) quantifies fire impacts. Lidar data integrated with GIS provides high-resolution 3D structural analysis to assess windthrow damage or pest-induced canopy mortality.

Spatial models predict insect outbreak susceptibility based on host species distribution, climate suitability, and forest connectivity (Hicke et al. 2016). For community adaptation planning, GIS transitions from assessment to active decision support:

Suitability Analysis for Adaptation Actions: GIS-based suitability modeling identifies optimal locations for interventions. This includes siting assisted migration trials (matching future climate niches with current species), prioritizing areas for fuel reduction treatments based on fire risk models and proximity to communities, identifying suitable sites for check dams or rainwater harvesting structures using watershed analysis (flow accumulation, slope), or delineating zones for mangrove restoration based on salinity, sediment dynamics, and historical cover (Rutherford et al. 2015; Barua et al. 2024).

Vulnerability Mapping for Targeted Support: Mapping community vulnerability (integrating forest dependency indices, socio-economic data, and climate exposure) allows governments and NGOs to prioritize resource allocation (e.g., early warning systems, alternative livelihood programs, health services) to the most at-risk populations, ensuring climate justice principles (Pandey et al. 2018).

Scenario Planning and Impact Forecasting: GIS facilitates exploring future scenarios by visualizing potential forest biome shifts under different emission pathways, projecting changes in NTFP availability, modeling future flood/inundation risks (e.g., sea-level rise in the Sundarbans), or simulating the effectiveness of different adaptation portfolios (e.g., impact of different mangrove buffer widths on storm surge attenuation) to inform long-term, resilient land-use planning (Sheppard 2012).

Participatory GIS (PGIS): Engaging local communities in mapping exercises using simplified GIS interfaces or sketch mapping integrated into GIS captures critical local knowledge on resource use, observed climate impacts, culturally significant sites, and preferred adaptation options. This fosters co-production of knowledge, enhances plan legitimacy, and empowers communities (McCall 2003).

While powerful, GIS-based approaches face challenges: data gaps (especially high-resolution climate and socio-economic data in developing regions), model uncertainties (particularly in downscaled climate projections and complex ecological responses), computational demands, and the need for capacity building to ensure local stakeholders can effectively utilize these tools (Table 9.2). Nevertheless, the ability of GIS to synthesize complex spatial information makes it an essential pillar for evidence-based climate adaptation in forest landscapes.

Table 9.2 GIS applications in supporting community adaptation planning

Adaptation planning need	GIS/geospatial tools and techniques	Community adaptation applications
Suitability analysis	Weighted overlay, multi-criteria evaluation, watershed analysis (flow accumulation, slope), habitat modeling	Siting rainwater structures, identifying mangrove restoration zones, planning fuel breaks, locating assisted migration trials
Resource vulnerability mapping	Spatial NTFP distribution models + climate projections, access/cost distance analysis	Mapping future NTFP availability, identifying communities most impacted by resource decline
Risk visualization and communication	3D visualization, WebGIS platforms, scenario mapping (sea-level rise, flood)	Communicating future risks to communities, participatory planning, evacuation route planning
Participatory planning (PGIS)	Community sketch mapping integrated into GIS, mobile data collection (ODK, KoBoToolbox), collaborative WebGIS	Documenting local knowledge, mapping community priorities, land-use conflict resolution, monitoring adaptation projects

9.4 A Case Study on Climate Change Impacts on Forest Landscapes and Community Adaptations

The Sundarbans, straddling the India-Bangladesh border and forming the world's largest contiguous mangrove forest, presents a stark and urgent case study of climate change impacts on a biodiverse forest landscape and the complex, often precarious, adaptation pathways of its dependent communities. This UNESCO World Heritage Site and Ramsar wetland, characterized by its intricate network of tidal waterways, mudflats, and islands dominated by salt-tolerant mangrove species like *Avicennia marina* and *Excoecaria agallocha*, faces a synergistic convergence of climate stressors: accelerated regional sea-level rise (3.2–5.8 mm/year, exceeding the global average), increased frequency and intensity of tropical cyclones (e.g., Sidr 2007, Aila 2009, Amphan 2020, Yaas 2021), rising temperatures, and altered freshwater flow from the Ganges-Padma system due to upstream diversions and changing precipitation patterns (Hazra et al. 2002; IPCC 2022). These forces drive salinity intrusion far inland, causing large-scale dieback of the economically and ecologically crucial Sundari tree (*H. fomes*), whose decline exceeds 20% in core areas, and a shift toward less biodiverse, monospecific stands of pioneer species. Satellite imagery analysis (Landsat, Sentinel-2) reveals significant loss of dense mangrove cover (estimated at 124 km^2 between 2000 and 2020 in the Indian sector) and erosion of critical islands like Ghoramara and Mousuni, directly undermining the natural storm surge buffering capacity of the mangroves, quantified as reducing wave energy by up to 90% (Dasgupta et al. 2022). Concurrently, ocean acidification and warming impact critical nursery habitats for commercially important fish and shellfish, with documented declines in Hilsa (*Tenualosa ilisha*) and Tiger Prawn (*Penaeus monodon*) catches, while increased human-tiger conflict occurs as habitat loss and saline intrusion push tigers closer to human settlements (Mitra 2019).

For the 5.4 million people residing in the Indian Sundarbans, primarily dependent on fishing, honey collection, small-scale agriculture, and NTFP harvesting, these ecological disruptions translate into profound livelihood insecurity, displacement, and heightened vulnerability. Cyclones and storm surges cause direct loss of life, livestock, and homes, while salinization of soil and groundwater renders vast tracts of agricultural land unproductive for traditional rice varieties, threatening food security. Declining fish and crab catches impoverish fishing communities, and the dangerous profession of wild honey collection from *Avicennia* forests becomes even more hazardous as floral patterns shift and tigers' range nearer. Women, who constitute a significant portion of agricultural labor and NTFP collectors, face disproportionate burdens in securing water and food amidst salinization (Ria 2025). Climate-induced displacement is a growing reality, with many becoming environmental refugees migrating to urban centers like Kolkata for precarious informal labor.

Community adaptations in the Sundarbans are diverse, often innovative, yet frequently constrained by poverty and institutional limitations: ***Agricultural Transformation***: Adoption of salt-tolerant rice varieties (e.g., *Nonabokra, Lunishree, Canning 7*) and brackish water aquaculture (e.g., *P. monodon* shrimp, *Scylla serrata* mud crab fattening) on former paddy land, though the latter can risk further salinization if unregulated (Dubey et al. 2017) (Fig. 9.2).

Livelihood Diversification: Increasing reliance on seasonal out-migration (men to construction, transport; women to domestic work), development of small-scale ecotourism, and NTFP value addition (e.g., processing *Hental* leaves for plates, honey cooperatives with improved safety and marketing) (Groth et al. 2023).

Structural and Ecological Protection: Raising homestead plinths using local materials, community-led mangrove regeneration (especially *Avicennia* and *Rhizophora* spp.) along vulnerable embankments, and participation in Early Warning Systems and cyclone shelters.

Water Management Innovations: Rainwater harvesting systems, community ponds protected from saline intrusion, and experimentation with floating gardens (*baira*) for vegetable cultivation in waterlogged areas. However, significant challenges persist. Maladaptation risks exist, such as poorly constructed embankments increasing vulnerability during breaches or unsustainable aquaculture practices. Governance failures, including weak implementation of Integrated Coastal Zone Management (ICZM) plans, inadequate maintenance of embankments, and limited inclusion of local knowledge in state-led adaptation projects, hinder effectiveness. Poverty and lack of access to finance restrict the scalability of successful local initiatives, while intersectional vulnerabilities (caste, gender, landlessness) mean the poorest often bear the brunt of impacts with the least adaptive capacity. The Sundarbans case underscores that effective adaptation requires hybrid approaches blending local ecological knowledge with scientific innovation, robust community institutions for

Fig. 9.2 Ecological and morphological features of mangrove habitats: **a** A panoramic view showing the interface between a dense mangrove canopy and an expansive, cracked mudflat during low tide; **b** A perspective of a narrow water channel or inlet where mangrove vegetation is beginning to colonize the silty banks; **c** Lush mangrove foliage extending over a brackish water body, illustrating the typical dense growth along river edges; **d** An elevated view showing a transition from dense forest to a structured pond, likely used for aquaculture or conservation research; **e** A row of young mangrove trees aligned along a sandy embankment, demonstrating their role in coastal protection; **f** A serene view of the mangrove treeline mirrored in the surface of a calm, inland pool; **g** A clear look at the tangled "prop roots" or stilt roots along a riverbank, which provide stability in soft, submerged soil; **h** A close-up of pneumatophores, the pencil-like aerial roots that rise from the mud to allow the trees to breathe in anaerobic conditions; **i** A narrow path cutting through the higher-ground vegetation, showing the diversity of plant life away from the immediate water's edge

collective action, equitable governance ensuring marginalized groups participate in decision-making, and significant investment in climate-resilient infrastructure and diversified livelihood options. It highlights mangroves not just as victims but as vital allies in adaptation, necessitating urgent global and national commitment to their protection and restoration as critical green infrastructure.

9.4.1 Materials and Methods

For analyzing long-term land cover dynamics and ecosystem characteristics, this study utilized Landsat satellite imagery, which has demonstrated proven effectiveness in time-series environmental monitoring (Vogelmann et al. 2012). Landsat Thematic Mapper (TM) and Operational Land Imager (OLI) datasets with 30-m spatial resolution were obtained from the United States Geological Survey (USGS) Earth Explorer portal. Imagery from 1990 and 2024 was selected to capture the temporal range of the study period. These datasets provided the foundation for Normalized Difference Vegetation Index (NDVI) analysis and subsequent land cover classification.

Secondary data sources and published literature were also incorporated to contextualize climate change impacts on forest landscape dynamics.

NDVI Analysis Section

NDVI was calculated from the selected satellite imagery to identify and map vegetation cover. This index is widely established in remote sensing applications for vegetation monitoring, as it compares near-infrared (NIR) and red band reflectance to assess vegetation health and vigor. The NDVI calculation follows the standard formula:

$$\text{NDVI} = \frac{(\text{NIR - Red})}{(\text{NIR + Red})} \tag{9.1}$$

NDVI values range from -1 to $+1$, where negative values typically indicate non-vegetated surfaces such as water bodies, snow, or bare soil. Values between 0 and 0.5 represent sparse to moderate vegetation cover, while values exceeding 0.5 indicate dense, healthy vegetation (Pettorelli et al. 2005; Xie et al. 2008). A threshold NDVI value was established to distinguish vegetated areas for subsequent forest and non-forest classification analysis.

Land Change Analysis

Transition intensity analysis was conducted using TerrSet software (formerly IDRISI) to quantify vegetation loss and urban expansion patterns. The Land Change

Modeler (LCM) module within TerrSet enabled pixel-by-pixel analysis of land cover transitions, revealing spatial patterns and rates of urban sprawl and deforestation processes. This integrated analytical approach facilitates comprehensive understanding of anthropogenic pressures affecting forest ecosystems within the context of rapid urbanization (Hasanuzzaman et al. 2022).

9.4.2 Results

Climate Change of Sundarbans, West Bengal

The Sundarbans, spanning the India-Bangladesh border, has experienced quantifiable and accelerating climate change impacts between 1990 and 2024, serving as a critical sentinel site for understanding deltaic vulnerability (Table 9.3). Accelerated Sea-Level Rise (SLR), significantly exceeding the global average, is the paramount driver. Tide gauge records (e.g., Sagar Island, Haldia) and satellite altimetry (TOPEX/Poseidon, Jason series) indicate a regional SLR rate of 3.2–8.0 mm/year during this period, compared to the global mean of ~ 3.3 mm/year (1993–2023), primarily attributed to thermal expansion, glacial melt, and local subsidence exacerbated by reduced sediment deposition due to upstream dams like Farakka (Hazra et al. 2002; IPCC 2022; PSMSL 2024). This has resulted in documented land loss exceeding 170 km^2 in the Indian Sundarbans since 1990, with islands like Ghoramara and Lohachara experiencing dramatic submergence and displacement of communities (Raha et al. 2012).

Concurrently, intensified tropical cyclones have become more frequent and destructive. Analysis of IMD data shows a notable increase in severe cyclonic storms (wind speed > 88 km/h) making landfall in the region post-2000 (e.g., Sidr 2007, Aila 2009, Bulbul 2019, Amphan 2020, Yaas 2021), characterized by higher storm surges (> 5 m during Amphan) and increased rainfall intensity, overwhelming traditional earthen embankments and causing catastrophic saline inundation (Mohapatra et al. 2025; IMD 2024). Salinity intrusion has penetrated 50–100 km inland since 1990, measured by increased conductivity in rivers and groundwater. Surface water salinity in the central Sundarbans increased by ~ 15–20% between 1990 and 2020, devastating the commercially and ecologically vital Sundari (*H. fomes*) mangroves, leading to > 25% decline in their density and a shift toward monospecific stands of less valuable pioneer species like *Avicennia* and *Excoecaria* (Mitra, 2019). Rising temperatures are unequivocal. Analysis of IMD station data reveals a significant warming trend, with mean annual temperatures increasing by ~ 0.8–1.2 °C between 1990 and 2023. Warming is more pronounced in the pre-monsoon months (March–May), exacerbating evapotranspiration and moisture stress.

While annual rainfall shows high inter-annual variability linked to the Indian Summer Monsoon, a trend toward more erratic precipitation is evident, with increased frequency of high-intensity rainfall events causing flash floods and landslides in the northern fringes, interspersed with prolonged dry spells (Ghosh et al.

Table 9.3 Documented climate trends in the Sundarbans (1990–2024)

Climate parameter	Observed change (1990–2024)	Primary data sources/ evidence	Key impacts
Sea-level rise	3.2–8.0 mm/ year (Regional > Global Avg)	Tide Gauges (Sagar, Haldia), Satellite Altimetry (TOPEX/ Jason), sediment core analysis	Land erosion (> 170 km^2 loss), Salinity intrusion, increased inundation
Tropical cyclones	Increased frequency and intensity of severe cyclonic storms (e.g., Amphan 2020—Cat 5)	IMD cyclone track and intensity records, satellite imagery (damage assessment)	Storm surge (> 5 m), embankment breaches, massive saline inundation, direct mortality
Surface water salinity	Increase of 15–20% in central Sundarbans	Long-term monitoring stations (WWF-I, JU), field conductivity measurements	Sundari Mangrove dieback (> 25% decline), shift to pioneer species, agricultural collapse
Mean temperature	Increase of ~ 0.8–1.2 °C (higher in pre-monsoon)	IMD station data (Canning, Sagar, Haldia), reanalysis products (ERA5)	Increased evapotranspiration, heat stress, altered species phenology
Precipitation	Increased variability; more intense rainfall events	IMD station data, TRMM/GPM satellite rainfall	Flash floods, landslides (North), prolonged dry spells, water stress
Sea surface temperature	Increase of ~ 0.7–1.0 °C (adjacent Bay of Bengal)	Satellite SST (AVHRR, MODIS), ship/Argo float data	Coral bleaching (offshore reefs), decline in fish stocks (Hilsa, Prawn), ocean acidification impacts

2015; Sahana et al. 2020). Ocean warming and acidification have also progressed, impacting critical marine biodiversity. Sea surface temperature (SST) in the Bay of Bengal adjacent to the Sundarbans has risen by ~ 0.7–1.0 °C since 1990, correlating with observed declines in key fishery resources like Hilsa shad (*T. ilisha*) and Tiger Prawn (*P. monodon*), disrupting the livelihoods of millions dependent on fishing and aquaculture (Mitra et al. 2009; IPCC 2022). These physical and ecological changes have driven profound socio-economic consequences (Table 9.4). Agricultural collapse is widespread in the central islands due to salinization, forcing abandonment of traditional rice farming. Brackish water aquaculture (shrimp, crab) expanded rapidly as an adaptation but brings risks of further salinization and disease outbreaks. Fisheries decline has impoverished communities, while increased human-tiger conflict reflects habitat compression and prey scarcity. Climate-induced displacement has accelerated, with migration to urban centers like Kolkata becoming a primary survival strategy, particularly for landless laborers (Harms 2020). Adaptation responses evolved significantly: early reliance on embankment repair post-cyclones (often maladaptive) shifted toward salt-tolerant rice cultivation (e.g., Canning 7), community-based mangrove restoration, raised

Table 9.4 Evolution of community impacts and adaptations (1990–2024)

Period	Dominant climate impacts	Primary community impacts	Key adaptation strategies
1990–2000	Moderate SLR, Cyclones (e.g., 1991 Bangladesh Cyclone), initial salinity rise	Localized land loss, early signs of Sundari stress, salt damage to peripheral rice fields	Embankment repair/maintenance, shift to limited brackish aquaculture, minor livelihood diversification
2000–2010	Accelerated SLR, Major Cyclones (Sidr 2007, Aila 2009), significant salinity intrusion	Widespread agricultural collapse, increased displacement, Major Sundari dieback, heightened human-tiger conflict	Expansion of aquaculture, Trials of salt-tolerant rice varieties, increased seasonal migration, community mangrove planting initiatives
2010–2020	Continued rapid SLR, intense cyclones (Bulbul 2019, Amphan 2020), severe salinization, warming	Large-scale land loss (island submergence), fisheries decline, water scarcity crises, health impacts (water-borne diseases)	Government/NGO-led EWS and cyclone shelters, scaling of salt-tolerant rice (e.g., Canning 7), homestead plinth raising, rainwater harvesting, formalization of NTFP cooperatives
2020–2024	Compound extremes (Cyclone Yaas 2021 + Tides), unabated SLR and salinity, heat stress	Deepening poverty traps, chronic displacement/ migration, ecosystem service degradation (protection, fisheries)	Hybrid adaptations (e.g., floating gardens), livelihood diversification beyond primary sectors, focus on climate-resilient infrastructure, increased advocacy/ recognition of climate refugees

homestead plinths, rainwater harvesting, and state-led Early Warning Systems and cyclone shelters. However, the pace and scale of these adaptations have been outpaced by the accelerating climate impacts, particularly for the most marginalized communities. The period 1990–2024 starkly illustrates how synergistic climate stressors-SLR, intensified cyclones, salinization, warming—are fundamentally reshaping this iconic deltaic ecosystem and challenging the resilience of its inhabitants.

Transformations of the Sundarbans Forest Landscape

The Sundarbans mangrove ecosystem has undergone profound, quantifiable transformations in its spatial extent, structure, composition, and health between 1990 and 2024, driven primarily by accelerating climate change impacts interacting with anthropogenic pressures (Table 9.5). ***Spatial Extent and Fragmentation***: Multi-temporal satellite imagery analysis (Landsat 5 TM, 7 ETM+, 8/9 OLI, Sentinel-2) reveals a net loss of approximately 124 km^2 (5.2%) of mangrove forest cover in the Indian sector during this period, primarily concentrated on the vulnerable sea-facing southern and southeastern margins of islands like Sagar, Ghoramara, Mousuni, and the central Sundarbans. This loss stems from direct erosion (estimated at > 170 km^2 total land area loss), submergence due to rising relative sea level, and cyclone-induced defoliation/uprooting followed by failed regeneration (Table 9.6). Concurrently, increased fragmentation is evident, with core forest areas shrinking and edge habitats expanding significantly, measured by rising edge density and patch number metrics, enhancing vulnerability to further degradation (Giri et al. 2007; Halder et al. 2021). ***Species Composition Shift***: The most ecologically significant change is the dramatic decline of the keystone species *H. fomes* (Sundari). Field inventories and spectral signature analysis indicate a > 70% reduction in Sundari density in core areas where it once dominated, primarily due to increased salinity stress exceeding its physiological tolerance (optimal < 15 ppt; now often > 20 ppt). This has triggered a successional shift toward monospecific or low-diversity stands of pioneer species like *A. marina* (gray mangrove), *Avicennia officinalis*, *E. agallocha* (blind-your-eye mangrove), and *Sonneratia apetala*. Areas once dominated by dense Sundari-Gewa (*Excoecaria*) forests now show significantly reduced canopy height, biomass, and structural complexity (Mitra, 2013; Sadeer et al. 2023).

Biophysical Degradation Indicators: Time-series analysis of vegetation indices derived from satellite data provides stark evidence of declining forest health. ***NDVI (Normalized Difference Vegetation Index)***: Shows significant negative trends,

Table 9.5 Quantified changes in Sundarbans forest landscape

Parameter	Observed change	Primary driver(s)	Measurement method	Key reference
Total Mangrove Area (Indian)	Net loss: ~ 124 km^2 (5.2%)	Erosion, submergence (SLR), cyclone damage	Landsat/sentinel time-series classification	Mondal et al. (2021)
Core Forest Area	Significant Decrease (> 20%)	Edge Erosion, Fragmentation	Landscape metrics (core area index)	Ghosh et al. (2025)
Heritiera fomes Density	> 70% decline in core habitats	Salinity intrusion (> 20 ppt)	Field inventories, spectral unmixing	Mitra (2013)
Canopy Height/ Biomass	Decrease (Shift to pioneer species)	Salinity, Cyclones, Reduced Growth	Lidar (limited), field plots, allometry	Shibu et al. (2022)
NDVI trend	Significant negative trend (central/East Sundarbans)	Salinity stress, moisture deficit	Landsat time-series analysis	Ghosh et al. (2025)
Fragmentation	Increased edge density, patch number	Erosion, localized dieback	Landscape metrics (FRAGSTATS)	Mondal et al. (2021)
Sediment accretion rate	Below SLR rate in most areas (net deficit)	Reduced river sediment load, SLR	Sediment cores, elevation surveys	Rahman et al. (2020)

Table 9.6 Spatial pattern of change in the Indian Sundarbans

Zone	Dominant landscape changes	Key species trends	Primary stressors
Southern Margins	Severe erosion, loss of front-line mangroves, flattening topography	Loss of *Sonneratia, Avicennia* pioneer colonization	Direct wave attack, cyclones, highest SLR exposure
Central Sundarbans	Major Sundari dieback, shift to *Avicennia/ Excoecaria*, fragmentation	> 70% *Heritiera* loss, proliferation of *Avicennia marina*	Highest salinity intrusion, reduced freshwater
Eastern Sundarbans	Similar to central, high salinity stress, some erosion	Strong decline in *Heritiera*, increase in *Ceriops*	Salinity, cyclones (e.g., Amphan landfall)
Western Sundarbans	Moderate erosion (active delta), lower salinity increase	Better persistence of mixed species (some Sundari), *Phoenix paludosa* decline	Active river channels, lower salinity
Northern Fringes	Historical conversion pressure, now flash flood/ erosion risk	Degraded patches, invasive species, freshwater species loss	Encroachment, erratic rainfall, salinity creep
Island Interiors	Relative stability, but fragmentation increasing	Slower decline of *Heritiera* (where present), dominance of *Excoecaria*	Salinity increase, groundwater salinization

particularly in the central and eastern Indian Sundarbans, indicating reduced photosynthetic activity and canopy greenness linked to salinity stress and storm damage. ***NDMI (Normalized Difference Moisture Index)***: Reveals increasing moisture stress (water deficit) across large swathes, correlating with reduced freshwater inflow and rising temperatures enhancing evapotranspiration. NBR (Normalized Burn Ratio): Applied to cyclone impact assessment, dNBR (differenced NBR) maps quantify severe canopy damage and slow recovery rates following major events like Aila (2009) and Amphan (2020), with some areas showing persistent degradation (Sahu et al. (2022).

Sediment Dynamics and Geomorphological Change: Reduced sediment load from the Ganges-Padma system (due to upstream dams like Farakka) coupled with rising sea levels has altered the critical sediment accretion-erosion balance. While some progradation occurs at river mouths (e.g., Hooghly), net sediment deficit prevails across most islands, hindering vertical accretion needed to keep pace with sea-level rise. This manifests in flatter island topographies, reduced elevation capital, and increased vulnerability to inundation, further stressing mangroves (Allison et al. 2003; Rahman et al. 2020).

Spatial Heterogeneity: Impacts are not uniform. The western Sundarbans (near the active Hooghly estuary) experiences lower salinity increases due to some freshwater flow, allowing slightly better persistence of mixed-species forests, though facing intense erosion. The central and eastern zones suffer the most severe salinity intrusion and Sundari dieback. The northern fringes face different pressures: conversion to agriculture/aquaculture historically and now increasing vulnerability to flash floods from erratic rainfall. Island interiors show some resilience, but edges are highly degraded (Ghosh et al. 2015, 2023). ***The "Greening" Paradox***: Some recent studies using high-resolution imagery note localized "greening" or increased NDVI in certain areas, often misinterpreted as recovery. This primarily reflects the proliferation of salt-tolerant, fast-growing pioneers like *Avicennia*, forming dense but ecologically simplified thickets, rather than a return to high-diversity, mature forest. It represents an ecosystem shift, not restoration to a previous state (Pineda et al. 2020).

The period 1990–2024 witnessed the Sundarbans transform from a system dominated by diverse, structurally complex forests featuring economically and ecologically vital Sundari toward a more fragmented, lower-stature, and salt-tolerant mangrove ecosystem. This shift, quantified through remote sensing and ground truthing, signifies a reduction in biodiversity, ecosystem services (timber, fisheries nursery habitat, carbon sequestration potential), and overall ecological resilience, posing immense challenges for both conservation and dependent communities.

Transformation of Forest to Non-forest in the Sundarbans (1990–2024)

Between 1990 and 2024, the Sundarbans region of West Bengal experienced a substantial loss of mangrove forest cover, with an estimated 1084.29 km^2 transformed into non-forest land. This transformation is a clear indicator of the combined impacts of anthropogenic pressures, climate change, and natural hazards in the world's largest contiguous mangrove ecosystem. The comparative satellite imagery from 1990 and 2024 (Fig. 9.3) illustrates a pronounced decline in dense mangrove patches, replaced increasingly by degraded scrubland, agricultural fields, aquaculture ponds, and settlements. Spatial analysis (Fig. 9.4) reveals that the loss is not uniformly distributed but is concentrated along the peripheries of inhabited islands and near tidal channels where erosion, embankment breaches, and saline water intrusion are prevalent. The magnitude of forest-to-non-forest conversion underscores the vulnerability of mangroves to both environmental stressors such as sea-level rise, increased frequency of cyclones, and salinity shifts and socio-economic drivers, including over-extraction of resources, land reclamation, and unsustainable aquaculture expansion. Such large-scale transformations not only threaten the ecological integrity and biodiversity of the Sundarbans but also compromise its role in coastal protection, carbon sequestration, and sustaining traditional livelihoods. The documented forest loss over the past three decades necessitates urgent policy interventions, community-based conservation strategies, and restoration programs to arrest further degradation and enhance the resilience of this UNESCO World Heritage Site (Uddin et al. 2013; Mukhopadhyay et al. 2021).

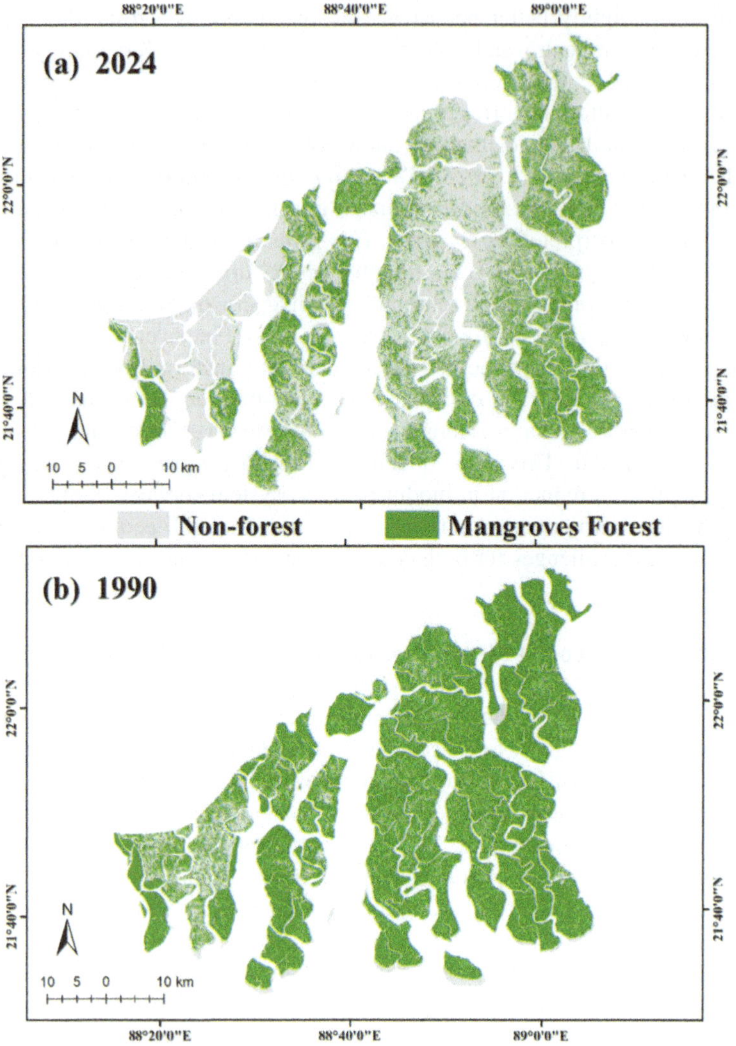

Fig. 9.3 Mangrove forest cover in the Sundarbans-**a** 1990 and **b** 2024

Climate Change and Its Impact on the Forest Landscape of the Sundarbans, West Bengal

From 1990 to 2024, the Sundarbans mangrove ecosystem has undergone profound ecological transformation, with climate change acting as the primary driver in reshaping its forest landscapes.

Accelerated sea-level rise (3.2–8.0 mm/year, exceeding the global mean) has resulted in over 170 km^2 of land loss, particularly along southern and southeastern

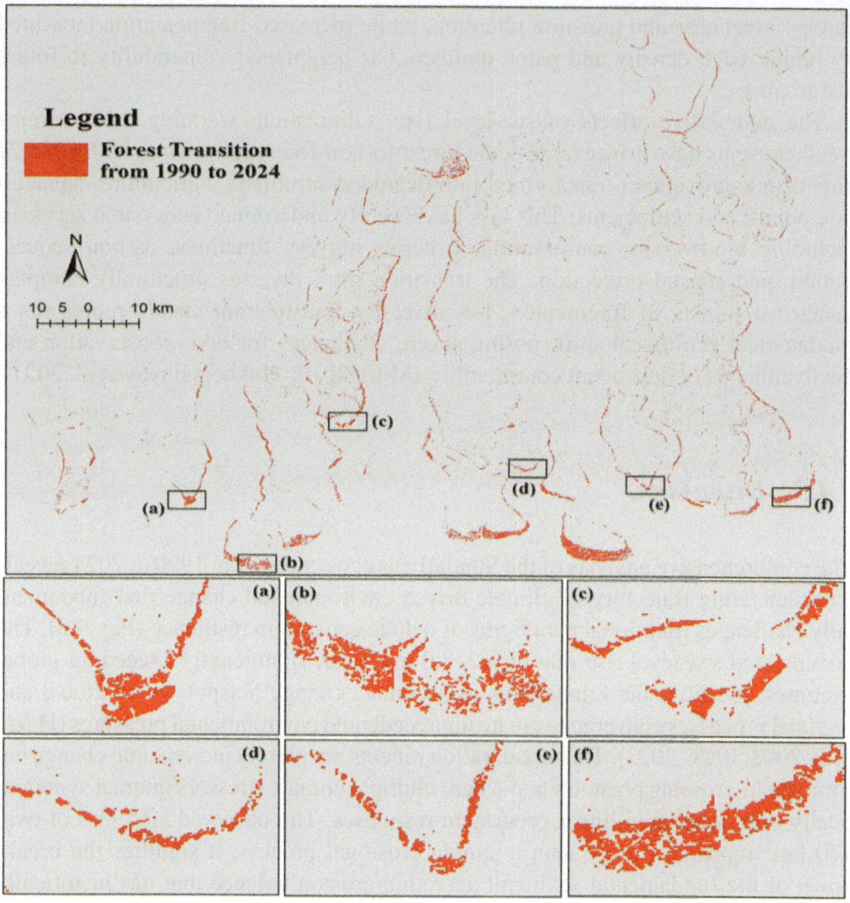

Fig. 9.4 Conversion of forest areas to non-forest land in the Sundarbans between 1990 and 2024

island margins, leading to direct erosion and submergence of mangrove stands. This hydrological stress is compounded by reduced sediment deposition from upstream damming, limiting the natural accretion needed for mangrove resilience. Increasing frequency and intensity of severe cyclones such as Aila (2009), Amphan (2020), and Yaas (2021) have caused widespread defoliation, uprooting, and saline inundation, with storm surges exceeding 5 m in some cases. This has amplified salinity intrusion by 15–20% in the central Sundarbans, pushing levels beyond the tolerance threshold of the keystone species *H. fomes* (Sundari). Consequently, Sundari density has declined by over 70%, triggering a compositional shift toward salt-tolerant pioneer species like *A. marina*, *E. agallocha*, and *S. apetala.* Warming trends (+0.8–1.2 °C in mean annual temperature) and erratic rainfall patterns have exacerbated evapotranspiration and moisture stress, further degrading mangrove health. Remote sensing indicators such as NDVI and NDMI show significant declines in canopy greenness and moisture retention, while increased fragmentation measured by higher edge density and patch numbers has heightened vulnerability to future disturbances.

The cumulative effects of sea-level rise, salinization, warming, and extreme weather events have driven large-scale forest-to-non-forest conversion, with 1084.29 km^2 of mangrove area transformed into degraded scrubland, agriculture, aquaculture ponds, and settlements. This loss has severely undermined ecosystem services, including biodiversity conservation, fisheries nursery functions, carbon sequestration, and coastal protection. The transition from diverse, structurally complex mangrove forests to fragmented, low-diversity, salt-tolerant stands represents a fundamental ecological shift, posing severe challenges for both conservation and the livelihoods of dependent communities (Mitra 2019; Mukhopadhyay et al. 2021).

9.4.3 *Discussion*

The comprehensive analysis of the Sundarbans ecosystem from 1990 to 2024 reveals an accelerating trajectory of climate-driven environmental change that fundamentally challenges traditional paradigms of deltaic ecosystem resilience (Fig. 9.5). The documented sea-level rise rate of 3.2–8.0 mm/year, significantly exceeding global averages, positions the Sundarbans as a climate change hotspot where global and regional processes converge to create unprecedented environmental pressures (Hazra et al. 2002; IPCC 2022). This acceleration reflects not merely incremental change but a threshold-crossing phenomenon where multiple climate stressors interact synergistically to produce non-linear ecosystem responses. The observed land loss of over 170 km^2 represents more than a simple erosional process; it signifies the breakdown of the fundamental sediment accretion-erosion balance that has historically sustained deltaic mangrove systems. The reduced sediment load from upstream damming, particularly the Farakka Barrage, has created a sediment deficit that leaves the ecosystem unable to maintain vertical accretion rates necessary to keep pace with rising seas (Allison et al. 2003; Rahman et al. 2020). This finding aligns with broader global patterns observed in other major deltas, including the Mississippi and Mekong, where anthropogenic modifications of river systems compound climate change impacts to accelerate land loss rates.

The > 70% decline in *H. fomes* (Sundari) density represents a classic example of ecological threshold crossing, where gradual environmental change suddenly precipitates dramatic ecosystem reorganization. The salinity tolerance threshold of Sundari (optimal < 15 ppt) has been exceeded across vast areas now experiencing > 20 ppt, triggering what appears to be an irreversible compositional shift toward pioneer species dominance (Mitra 2019; Mukhopadhyay et al. 2021). This transformation exemplifies the concept of alternative stable states in ecology, where ecosystems can exist in multiple configurations under the same environmental conditions, but transitions between states may be difficult or impossible to reverse. The proliferation of *A. marina*, *E. agallocha*, and *S. apetala* in areas previously dominated by diverse,

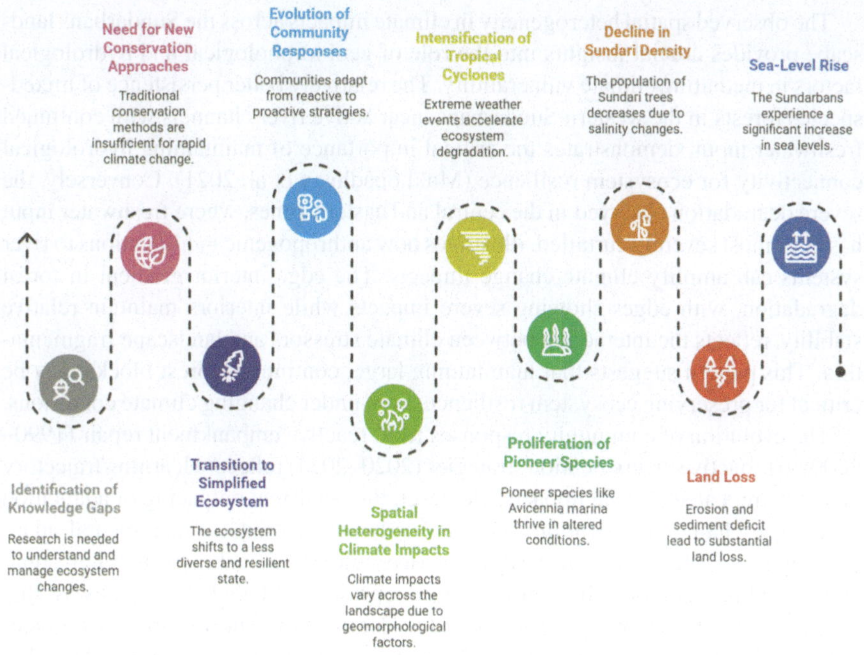

Fig. 9.5 Symmetrical flowchart of environmental changes in the Sundarbans ecosystem

structurally complex forests represents not recovery but rather ecosystem simplification a shift toward a more stress-tolerant but functionally impoverished system. This pattern mirrors observations from other climate-stressed mangrove systems globally, including those in Florida Bay and northern Australia, where increasing salinity and temperature stress have similarly driven compositional shifts toward halophytic pioneer species.

The intensification of tropical cyclones, particularly the unprecedented Category 5 Cyclone Amphan in 2020, demonstrates how extreme events can serve as ecological catalysts that rapidly accelerate ongoing gradual changes. The > 5-m storm surge heights observed during recent cyclones exceed the design capacity of traditional coastal protection infrastructure and overwhelm the natural buffering capacity of degraded mangrove systems (Mukhopadhyay et al. 2021; IMD 2024). This creates a positive feedback loop where climate-damaged ecosystems become increasingly vulnerable to subsequent climate impacts. The temporal clustering of severe cyclones (Sidr 2007, Aila 2009, Bulbul 2019, Amphan 2020, Yaas 2021) illustrates the emerging reality of compound climate extremes, where recovery periods between disturbances become insufficient for ecosystem regeneration. The persistent degradation observed in dNBR (differenced Normalized Burn Ratio) analyses following major cyclones suggests that recovery trajectories have been fundamentally altered, with some areas trapped in degraded states unable to return to pre-disturbance conditions (Shibu et al. 2022).

The observed spatial heterogeneity in climate impacts across the Sundarbans landscape provides crucial insights into the role of geomorphological and hydrological factors in mediating climate vulnerability. The relatively better persistence of mixed-species forests in the western Sundarbans, near active river channels with continued freshwater input, demonstrates the critical importance of maintaining hydrological connectivity for ecosystem resilience (Mukhopadhyay et al. 2021). Conversely, the severe degradation observed in the central and eastern zones, where freshwater input has been most severely curtailed, illustrates how anthropogenic modifications to river systems can amplify climate change impacts. The edge-interior gradient in forest degradation, with edges showing severe impacts while interiors maintain relative stability, reflects the interaction between climate stressors and landscape fragmentation. This pattern suggests that maintaining large, continuous forest blocks may be critical for preserving ecosystem resilience, even under changing climate conditions.

The evolution of community responses from reactive embankment repair (1990–2000) to proactive diversification strategies (2020–2024) reflects a learning trajectory driven by repeated climate impacts. However, the persistent outpacing of adaptation by climate change acceleration suggests fundamental limits to incremental adaptive capacity. The shift from traditional rice cultivation to brackish water aquaculture, while economically rational, may represent a maladaptive pathway that further accelerates salinization and ecosystem degradation. The documented climate-induced displacement patterns align with broader migration trends observed in other climate-vulnerable deltaic regions, including the Nile Delta and Mekong Delta, suggesting common socio-ecological vulnerabilities across deltaic systems globally. The emergence of "climate refugees" from the Sundarbans highlights the inadequacy of existing institutional frameworks for addressing climate-driven displacement.

The Sundarbans case provides a sobering preview of potential futures for mangrove ecosystems globally under accelerating climate change. The transition from diverse, high-biomass forests to simplified, low-stature pioneer stands represents a fundamental shift in ecosystem function that has cascading implications for global carbon cycling, coastal protection, and biodiversity conservation. The estimated loss of 1084.29 km^2 of mangrove forest between 1990 and 2024 represents not only local ecological degradation but also a significant contribution to global mangrove loss rates. The "greening paradox" observed in some areas where increased NDVI values reflect pioneer species proliferation rather than ecosystem recovery highlights the importance of using multiple indicators to assess ecosystem health and the dangers of misinterpreting remote sensing data without ground-truth validation (Samanta et al. 2021).

The Sundarbans experience suggests that traditional conservation approaches focused on preserving existing ecosystem states may be insufficient under rapid climate change. Instead, management strategies may need to focus on managing ecosystem transformations to maintain essential functions while accepting compositional and structural changes. This "threshold management" approach recognizes that some changes may be irreversible while focusing on preventing cascading failures that could lead to complete ecosystem collapse. The success of salt-tolerant rice varieties (e.g., Canning 7) and community-based mangrove restoration initiatives suggests that hybrid approaches combining technological innovation with ecosystem-based adaptation may offer pathways forward, even under continued climate change acceleration.

The Sundarbans case highlights critical knowledge gaps that limit our ability to predict and manage ecosystem responses to accelerating climate change. Long-term monitoring networks need to be expanded to capture threshold effects and early warning signals of impending state changes. Experimental approaches are needed to test the effectiveness of novel restoration techniques under projected future climate conditions. Integration of indigenous and scientific knowledge systems appears crucial for developing locally appropriate adaptation strategies that can operate across the full spectrum of social-ecological scales. The documented evolution of community adaptation strategies provides valuable insights that could inform climate adaptation planning in other vulnerable deltaic regions globally. The Sundarbans transformation from 1990 to 2024 thus represents both a regional ecological tragedy and a global climate change laboratory, providing essential insights into the dynamics of ecosystem transformation under accelerating environmental change. Understanding these processes is crucial not only for the future of the Sundarbans but for anticipating and managing climate change impacts on deltaic ecosystems worldwide.

9.4.4 Conclusion

The comprehensive analysis of the Sundarbans mangrove ecosystem from 1990 to 2024 documents unprecedented climate-driven transformation in one of the world's most significant deltaic systems. This study reveals fundamental reorganization of ecosystem structure, function, and socio-ecological relationships with profound implications for understanding global coastal ecosystem vulnerability under accelerating climate change. The scale and pace of change far exceed previous projections. Regional sea-level rise of 3.2–8.0 mm/year, substantially exceeding global averages, coupled with land loss exceeding 170 km^2 and forest cover reduction of 1084.29 km^2, represents one of the most rapid landscape transformations in contemporary mangrove science. The > 70% decline in *H. fomes* density illustrates that climate impacts involve threshold effects fundamentally altering ecosystem identity within human timescales. Multiple climate stressors interact synergistically to produce impacts exceeding individual stresses. The convergence of accelerated sea-level rise, intensified cyclones, increased salinity intrusion (15–20% increase), rising temperatures (0.8–1.2 °C), and altered precipitation created ecological disruptions overwhelming traditional resilience mechanisms. Temporal clustering of severe cyclones (Sidr 2007, Aila 2009, Bulbul 2019, Amphan 2020, Yaas 2021) demonstrates compound climate extremes where recovery periods become insufficient for ecosystem regeneration. Pronounced spatial variability provides insights into vulnerability factors. Better persistence of mixed-species forests in western Sundarbans, maintained by freshwater inputs, contrasts with severe degradation in central and eastern zones where connectivity was compromised. This demonstrates hydrological connectivity's critical role in ecosystem resilience. Community responses evolved from reactive strategies (1990–2000) to sophisticated adaptation measures (2020–2024), illustrating both human adaptability and incremental adaptation limits. Despite innovations in salt-tolerant agriculture and infrastructure improvements, climate change acceleration consistently outpaced adaptation efforts. Climate-induced displacement emerged as a primary survival strategy, highlighting place-based adaptation breakdown. The Sundarbans provides a preview of deltaic mangrove futures under accelerating climate change. Ecosystem shifts from diverse, high-biomass forests to simplified pioneer stands represent fundamental functional changes with cascading implications for carbon cycling, coastal protection, and biodiversity conservation. The "greening paradox" where increased vegetation indices reflect pioneer proliferation rather than recovery emphasizes ground-truth validation importance. Traditional conservation approaches focused on preserving existing states appear insufficient under rapid climate change. Management strategies must embrace "threshold management" maintaining essential functions while accepting compositional changes. Hybrid adaptation strategies combining technological innovation with ecosystem-based approaches suggest transformation pathways, though local measures may be insufficient without addressing root causes through global emissions reduction. The trajectory suggests the Sundarbans may have crossed critical thresholds beyond which historical states cannot be restored, necessitating transformation-oriented management guiding change toward desired futures rather than recreating past conditions. Novel ecosystems appear increasingly likely under continued acceleration. Critical knowledge gaps limit prediction and management capabilities. Long-term monitoring networks need expansion to capture threshold effects and early warnings. Integration of indigenous and scientific knowledge systems emerges as crucial for locally appropriate strategies. The Sundarbans transformation represents both regional tragedy and global climate laboratory, serving as a sentinel for understanding deltaic system impacts worldwide. Lessons learned threshold effects, synergistic stressors, spatial heterogeneity, and adaptation limits have direct relevance for managing climate impacts on coastal ecosystems globally. The period 1990–2024 represents a critical juncture offering sobering warnings and valuable insights for ecosystem management under unprecedented environmental change.

References

Allison MA, Khan SR, Goodbred SL Jr, Kuehl SA (2003) Stratigraphic evolution of the late Holocene Ganges-Brahmaputra lower delta plain. Sed Geol 155(3–4):317–342

Abatzoglou JT, Williams AP (2016) Impact of anthropogenic climate change on wildfire across western US forests. Proc Natl Acad Sci 113(42):11770–11775

Adger WN, Barnett J, Brown K, Marshall N, O'brien K (2013) Cultural dimensions of climate change impacts and adaptation. Nat Clim Change 3(2):112–117

Allen CD, Macalady AK, Chenchouni H, Bachelet D, McDowell N, Vennetier M et al (2010) A global overview of drought and heat-induced tree mortality reveals emerging climate change risks for forests. For Ecol Manage 259(4):660–684

Basu JP (2020) Climate change vulnerability and communities in agro-climatic regions of West Bengal, India: theory and practice. Springer Nature

Barua P, Chowdhury SS, Mitra A (2024) Review on conservation approach study of largest mangrove ecosystem as the mitigation of climate change. J Clim Change 10(3):31–42

Berkes F (2017) Sacred ecology. Routledge

Chhatre A, Agrawal A (2009) Trade-offs and synergies between carbon storage and livelihood benefits from forest commons. Proc Natl Acad Sci 106(42):17667–17670

Das GK (2021) Forests and forestry of West Bengal: survey and analysis. Springer Nature

DasGupta R, Shaw R (2013) Cumulative impacts of human interventions and climate change on mangrove ecosystems of South and Southeast Asia: an overview. J Ecosyst 2013(1):379429

Dasgupta R, Hashimoto S, Basu M, Okuro T, Johnson BA, Kumar P, Dhyani S (2022) Spatial characterization of non-material values across multiple coastal production landscapes in the Indian Sundarban delta. Sustain Sci 17(3):725–738

Dubey SK, Trivedi RK, Chand BK, Mandal B, Rout SK (2017) Farmers' perceptions of climate change, impacts on freshwater aquaculture and adaptation strategies in climatic change hotspots: a case of the Indian Sundarban delta. Environ Develop 21:38–51

Fernández-Llamazares Á, Lepofsky D, Lertzman K, Armstrong CG, Brondizio ES (2021)

Ford JD (2012) Indigenous health and climate change. Am J Public Health 102(7):1260–1266

Ghosh A, Schmidt S, Fickert T, Nüsser M (2015) The Indian Sundarban mangrove forests: history, utilization, conservation strategies and local perception. Diversity 7(2):149–169

Ghosh A, Kumar S, Das J (2023) Impact of leachate and landfill gas on the ecosystem and health: Research trends and the way forward towards sustainability. J Environ Manag 336:117708

Ghosh T, Hazra S, Malick K, Talukdar T (2025) Monitoring ground deformation in recent years in Western Ganga delta India, using persistent scatterer interferometry on multi-temporal ALOS-2 L band imaging radar data. Nat Hazards 1–31

Giri C, Pengra B, Zhu Z, Singh A, Tieszen LL (2007) Monitoring mangrove forest dynamics of the Sundarbans in Bangladesh and India using multi-temporal satellite data from 1973 to 2000. Estuar Coast Shelf Sci 73(1–2):91–100

Goetz SJ, Sun M, Baccini A, Beck PS (2010) Synergistic use of spaceborne lidar and optical imagery for assessing forest disturbance: an Alaska case study. J Geophys Res Biogeosci 115(G2)

Groth J, Seppelt R, Sakdapolrak P, Senbeta F, Hermans K (2023) Why smallholders stop engaging in forest activities: the role of in-migration in livelihood transitions in forested landscapes of southwestern Ethiopia. Ecol Soc 28(1)

Halder S, Samanta K, Das S, Pathak D (2021) Monitoring the inter-decade spatial–temporal dynamics of the Sundarban mangrove forest of India from 1990 to 2019. Reg Stud Mar Sci 44:101718

Hasanuzzaman M, Gayen A, Mafizul Haque S, Shit PK (2022) Spatial modeling of river bank shifting and associated LULC changes of the Kaljani River in Himalayan foothills. Stoch Env Res Risk Assess 36(2):563–582

Harms A (2020) Under the climate radar: disaster and displacement in the Bengal Delta. In: Climate justice and migration: mobility, development, and displacement in the global south. Heinrich Böll Stiftung, pp 150–155

Hazra S, Ghosh T, DasGupta R, Sen G (2002) Sea level and associated changes in the Sundarbans. Sci Cult 68(9/12):309–321

Hicke JA, Meddens AJ, Kolden CA (2016) Recent tree mortality in the western United States from bark beetles and forest fires. Forest Sci 62(2):141–153

IMD (India Meteorological Department) (2024) Cyclone eAtlas—IMD. (Data portal for historical cyclone tracks and intensity)

IPCC (2022) Climate Change 2022: impacts, adaptation and vulnerability. In: Pörtner H-O et al (eds) Contribution of working group II to the sixth assessment report of the intergovernmental panel on climate change. Cambridge University Press

Kofinas GP, Chapin III FS (2009) Sustaining livelihoods and human well-being during social-ecological change. In: Principles of ecosystem stewardship: resilience-based natural resource management in a changing world. Springer New York, pp 55–75

Kumar R, Saikia P (2020) Forest resources of Jharkhand, Eastern India: socio-economic and bio-ecological perspectives. Socio-economic and Eco-biological dimensions in resource use and conservation: strategies for sustainability. Springer International Publishing, Cham, pp 61–101

Malik A, Singh H, Kumar K, Kumari G, Sojitra A (2024) Understanding the Impact of Climate Change and environments on urban forests: insights from morphological, physiological, and molecular perspectives. In: Urban forests, climate change and environmental pollution: physio-biochemical and molecular perspectives to enhance urban resilience. Springer Nature Switzerland, Cham, pp 161–183

Metcalf EC, Mohr JJ, Yung L, Metcalf P, Craig D (2015) The role of trust in restoration success: public engagement and temporal and spatial scale in a complex social-ecological system. Restor Ecol 23(3):315–324

McDowell NG, Allen CD, Anderson-Teixeira K, Aukema BH, Bond-Lamberty B, Chini L et al (2020) Pervasive shifts in forest dynamics in a changing world. Science 368(6494):eaaz9463

McCall MK (2003) Seeking good governance in participatory-GIS: a review of processes and governance dimensions in applying GIS to participatory spatial planning. Habitat Int 27(4):549–573

Millar CI, Stephenson NL, Stephens SL (2007) Climate change and forests of the future: managing in the face of uncertainty. Ecol Appl 17(8):2145–2151

Mitra A (2013) Sensitivity of mangrove ecosystem to changing climate (vol 62). Springer, New Delhi, pp 143–157

Mitra A, Gangopadhyay A, Dube A, Schmidt AC, Banerjee K (2009) Observed changes in water mass properties in the Indian Sundarbans (northwestern Bay of Bengal) during 1980–2007. Curr Sci 1445–1452

Mitra J (2019) Entrepreneurship, innovation and regional development: an introduction. Routledge

Mohapatra M, Sharma M, Pattanaik DR (2025) Evolution of cyclone monitoring and forecasting system in India: a review. Mausam 76(1):103–124

Mondal I, Thakur S, Ghosh P, De TK (2021) Assessing the impacts of global sea level rise (SLR) on the mangrove forests of Indian Sundarbans using geospatial technology. Geographic information science for land resource management (pp 209–227)

Mukhopadhyay A, Mitra D, Hazra S (eds) (2021) Sundarbans Mangrove systems: a geo-informatics approach. CRC Press

Nagel LM, Palik BJ, Battaglia MA, D'Amato AW, Guldin JM, Swanston CW et al (2017) Adaptive silviculture for climate change: a national experiment in manager-scientist partnerships to apply an adaptation framework. J Forest 115(3):167–178

O'Brien K, Leichenko R, Kelkar U, Venema H, Aandahl G, Tompkins H et al (2004) Mapping vulnerability to multiple stressors: climate change and globalization in India. Glob Environ Chang 14(4):303–313

Pandey R, Kumar P, Archie KM, Gupta AK, Joshi PK, Valente D, Petrosillo I (2018) Climate change adaptation in the western-Himalayas: household level perspectives on impacts and barriers. Ecol Ind 84:27–37

Pettorelli N, Vik JO, Mysterud A, Gaillard JM, Tucker CJ, Stenseth NC (2005) Using the satellite-derived NDVI to assess ecological responses to environmental change. Trends Ecol Evol 20(9):503–510

Pineda M, Barón M, Pérez-Bueno ML (2020) Thermal imaging for plant stress detection and phenotyping. Remote Sens 13(1):68

PSMSL (Permanent Service for Mean Sea Level) (2024) Tide Gauge Data [Data for Haldia/Sagar Island Stations]. Retrieved July 26, 2025, from https://psmsl.org/data/obtaining/

Raffa KF, Aukema BH, Bentz BJ, Carroll AL, Hicke JA, Turner MG, Romme WH (2008) Cross-scale drivers of natural disturbances prone to anthropogenic amplification: the dynamics of bark beetle eruptions. Bioscience 58(6):501–517

Raha A, Das S, Banerjee K, Mitra A (2012) Climate change impacts on Indian Sunderbans: a time series analysis (1924–2008). Biodivers Conserv 21(5):1289–1307

Rahman MM, Ghosh T, Salehin M, Ghosh A, Haque A, Hossain MA et al (2020) Ganges-Brahmaputra-Meghna delta, Bangladesh and India: a transnational mega-delta. Deltas Anthropocene 23–51

Ravindranath NH, Rao S, Sharma N, Nair M, Gopalakrishnan R, Rao AS et al (2011) Climate change vulnerability profiles for North East India. Curr Sci 384–394

Rutherford J, Kobryn H, Newsome D (2015) A case study in the evaluation of geotourism potential through geographic information systems: application in a geology-rich island tourism hotspot. Curr Issue Tour 18(3):267–285

Ria A (2025) A feminist perspective of the differentiated impacts of climate change, adaptation and women's roles in coastal agriculture in Bangladesh (Doctoral dissertation, University of Manitoba)

Sadeer NB, Zengin G, Mahomoodally MF (2023) Biotechnological applications of mangrove plants and their isolated compounds in medicine-a mechanistic overview. Crit Rev Biotechnol 43(3):393–414

Sahu TK, Shukla PC, Belgiorno G, Maurya RK (2022) Alcohols as alternative fuels in compression ignition engines for sustainable transportation: A review. Energy Sources, Part A: Recovery, Util Environ Eff 44(4):8736–8759

Sahana M, Rehman S, Patel PP, Dou J, Hong H, Sajjad H (2020) Assessing the degree of soil salinity in the Indian Sundarban Biosphere Reserve using measured soil electrical conductivity and remote sensing data–derived salinity indices. Arab J Geosci 13(24):1289

Samanta S, Hazra S, Mondal PP, Chanda A, Giri S, French JR, Nicholls RJ (2021) Assessment and attribution of mangrove Forest changes in the Indian Sundarbans from 2000 to 2020. Remote Sens 13(24):4957

Seidl R, Thom D, Kautz M, Martin-Benito D, Peltoniemi M, Vacchiano G et al (2017) Forest disturbances under climate change. Nat Clim Chang 7(6):395–402

Sheppard SR (2012) Visualizing climate change: a guide to visual communication of climate change and developing local solutions. Routledge

Shibu AV, Kurup BM, Harikrishnan M, Boopendranath MR (2022) Impact of climate change on hydrological cycle, ecosystem, fisheries and food security

Tambe S, Kharel G, Arrawatia ML, Kulkarni H, Mahamuni K, Ganeriwala AK (2012) Reviving dying springs: climate change adaptation experiments from the Sikkim Himalaya. Mt Res Dev 32(1):62–72

Tiwary M (2019) Participatory forest policies and politics in India: joint forest management institutions in Jharkhand and West Bengal. Routledge

Uddin MS, Van Steveninck EDR, Stuip M, Shah MAR (2013) Economic valuation of provisioning and cultural services of a protected mangrove ecosystem: a case study on Sundarbans Reserve Forest, Bangladesh. Ecosyst Serv 5:88–93

Vogelmann JE, Xian G, Homer C, Tolk B (2012) Monitoring gradual ecosystem change using Landsat time series analyses: case studies in selected forest and rangeland ecosystems. Remote Sens Environ 122:92–105

Wang X, Thompson DK, Marshall GA, Tymstra C, Carr R, Flannigan MD (2015) Increasing frequency of extreme fire weather in Canada with climate change. Clim Change 130(4):573–586

Whyte K (2017) Indigenous climate change studies: Indigenizing futures, decolonizing the Anthropocene. Engl Lang Notes 55(1):153–162

Xie Y, Sha Z, Yu M (2008) Remote sensing imagery in vegetation mapping: a review. J Plant Ecol 1(1):9–23

Zhu Z (2017) Change detection using landsat time series: a review of frequencies, preprocessing, algorithms, and applications. ISPRS J Photogramm Remote Sens 130:370–384

Chapter 10
Conclusion: Toward Integrated Forest Landscape Governance in West Bengal

Abstract West Bengal's forest landscapes represent critically contested social-ecological systems where centuries of colonial dispossession, post-independence centralization, accelerating fragmentation, urban encroachment, and climate variability have simultaneously degraded ecological integrity and deepened the livelihood vulnerabilities of millions of forest-dependent tribal and marginalized communities. Despite pioneering participatory governance experiments including the landmark Arabari Joint Forest Management initiative, persistent structural inequalities, bureaucratic dominance, elite capture, and inadequate tenure security continue undermining sustainable forest stewardship across this ecologically diverse state. This study synthesizes multi-dimensional evidence across historical governance transitions, spatial fragmentation dynamics, carbon stock heterogeneity, ecosystem service valuation, livelihood dependencies, and participatory management outcomes to propose an integrated social-ecological governance framework for West Bengal's forest landscapes. A convergent mixed-methods approach was employed, integrating multi-temporal Landsat/Sentinel-2 satellite imagery, GIS-based fragmentation metrics, Random Forest machine learning classification, InVEST ecosystem service modeling, allometric biomass estimation, stratified household surveys, participatory rural appraisal, Livelihood Dependency Index computation, and institutional policy analysis across ecologically and socially heterogeneous districts including Jhargram, Bankura, and the Sundarbans. Forest cover declined by 30.2% between 1995 and 2025, patch numbers increased by 340%, and total ecosystem service value contracted by 49% (from $775.11 to $395.24/ha annually). Non-timber forest products constituted 47% of tribal household income, yet forest proximity paradoxically correlated with lower living standards. Joint Forest Management improved carbon sequestration by 15–25% but exhibited severe gender exclusion, with female decision-making participation below 2%. Achieving ecologically resilient and socially equitable forest futures demands polycentric governance reforms: genuine Forest Rights Act implementation, mandatory equity-based representation within Forest Protection Committees, landscape-scale corridor conservation, climate-adaptive management strategies, and systematic integration of traditional ecological knowledge offering transferable policy frameworks for tropical forest governance across the Global South.

P. K. Shit and M. Hasanuzzaman, *Forest Landscape Dynamics in West Bengal*,
Advances in Asian Human-Environmental Research,
https://doi.org/10.1007/978-3-032-23700-2_10

Keywords Integrated forest governance · Social-ecological systems · Forest fragmentation · Ecosystem services · Joint Forest Management · Carbon sequestration · Livelihood dependency · Traditional ecological knowledge · Climate adaptation · West Bengal

10.1 Synthesis of Findings and Pathways Forward

The journey through West Bengal's Forest landscapes from their historical transformations to contemporary ecological challenges and community adaptations reveals a complex tapestry of interactions between human societies and natural systems. This volume has examined how forests, once governed through indigenous stewardship rooted in traditional ecological knowledge, have undergone profound structural changes through colonial exploitation, post-independence centralization, and recent experiments in participatory management. The narrative that emerges is neither one of simple degradation nor straightforward recovery, but rather a continuous negotiation between competing visions of forest use, conservation imperatives, and livelihood dependencies.

10.2 Historical Legacies and Contemporary Realities

The historical analysis presented in Chap. 2 demonstrates that contemporary forest conditions cannot be understood without acknowledging the deep structural transformations initiated during the colonial period. The conversion of community-managed resources into state property through the Forest Acts of 1865 and 1878 fundamentally altered the relationship between forests and people. Indigenous management systems that had maintained ecological balance for centuries through sacred groves, rotational harvesting, and customary regulatory frameworks were systematically dismantled and replaced with extractive forestry focused on revenue generation. This transition established patterns of centralized control and marginalized traditional knowledge, effects that persist despite decades of independence and policy reform.

The post-independence period initially reinforced these colonial structures, with the retention of the Indian Forest Act of 1927 and further centralization through state acquisition acts. However, the emergence of Joint Forest Management in the 1970s and 1980s marked a conceptual breakthrough, recognizing that sustainable forest governance requires active community participation. The Arabari experiment, detailed in Chap. 8, demonstrated that when communities receive genuine rights and benefits, forest regeneration can occur even in severely degraded landscapes. Yet the implementation challenges bureaucratic dominance, elite capture, gender exclusion, and incomplete devolution of authority underscore that institutional change alone is insufficient without fundamental shifts in power relations.

10.3 Spatial Dynamics of Forest Degradation and Fragmentation

The geospatial analyses presented throughout this volume reveal the profound spatial dimensions of forest change in West Bengal. Chap. 3's examination of forest fragmentation in Jhargram district exemplifies broader patterns across the state, where continuous forest cover has been progressively subdivided into smaller, isolated patches. The 30.2% decline in forest cover between 1995 and 2025, coupled with a 340% increase in patch numbers, represents more than mere statistics it signifies fundamental alterations in ecosystem structure and function. The particularly severe reduction in large core forests, which declined by 34.9%, threatens the very foundation of biodiversity conservation and genetic connectivity.

These fragmentation patterns are not uniform across the landscape but reflect specific drivers operating at different spatial scales. Agricultural expansion remains the dominant force, accounting for 62% of fragmentation, followed by infrastructure development at 27%. The case of Siliguri, explored in Chap. 5, illustrates how urban expansion can rapidly transform peri-urban-forest landscapes, with built-up area increasing by 253% between 1990 and 2025. Such expansion disrupts critical wildlife corridors, intensifies human-wildlife conflict, and diminishes the capacity of forests to provide essential ecosystem services.

The spatial heterogeneity of forest degradation demands spatially differentiated management strategies. Fragmentation in the southwestern lateritic tracts occurs through perforation driven by small-scale clearing and fuelwood extraction, while northern districts experience edge-dominated fragmentation from road construction and settlement expansion. The Sundarbans face unique challenges from hydrological disruption and aquaculture encroachment. Recognition of these distinct spatial patterns is essential for designing effective conservation interventions that address local drivers while maintaining landscape-scale connectivity.

10.4 Carbon Dynamics and Climate Change Mitigation

Chapter 4's analysis of forest biomass and carbon stocks positions West Bengal's forests within the global context of climate change mitigation. With forests storing approximately 861 gigatons of carbon globally nearly 45% of all terrestrial carbon the state's 16,902 km^2 of forest represent significant, though heterogeneous, carbon reservoirs. The Sundarbans emerge as exceptional carbon sinks with ecosystem carbon stocks of 400–600 Mg C per hectare, while sal-dominated forests store 40–90 Mg C per hectare, and montane forests of Darjeeling hold 120–250 Mg per hectare of biomass.

The Jangal Mahal case study provides granular insights into carbon stock variability, revealing above-ground biomass ranging from 37.4 to 268.6 tons per hectare, with native Shorea robusta forests demonstrating superior carbon efficiency compared to monoculture plantations. This finding carries important implications for forest management strategies the prioritization of natural forest conservation over plantation forestry not only preserves biodiversity but also optimizes carbon sequestration potential. The observation that Joint Forest Management programs have improved sequestration rates by 15–25% in community-managed forests suggests that participatory governance can simultaneously advance climate mitigation and community welfare objectives.

However, ongoing anthropogenic pressures and annual deforestation rates of 0.2–0.5% continue to reduce forest carbon density. The challenge ahead involves reversing these trends through integrated strategies that combine protection of remaining high-carbon forests, restoration of degraded landscapes, and enhancement of carbon stocks through sustainable management practices. The integration of remote sensing technologies with machine learning algorithms, achieving up to 85% accuracy in carbon mapping, provides powerful tools for monitoring progress and identifying priority areas for intervention.

10.5 Biodiversity Conservation and Ecosystem Service Provision

The intricate relationships between biodiversity and ecosystem services examined in Chap. 6 underscore that forest value extends far beyond timber production or carbon storage. Forests provide provisioning services including non-timber forest products essential for rural livelihoods, regulating services such as water purification and climate moderation, cultural services encompassing recreation and spiritual values, and supporting services like nutrient cycling and soil formation. The Bankura case study quantifies this multifunctionality, revealing that total ecosystem service value declined by 49% between 1990 and 2024, from 775.11 to 395.24 US dollars per hectare annually.

This dramatic reduction in ecosystem service provision reflects both the extent of forest degradation and the consequences of fragmentation. Edge and patch forests exhibited the steepest declines of 58–72%, while large core forests retained relatively higher stability (Fig. 10.1). The spatial pattern of ecosystem service loss varied across community development blocks, with Sonamukhi, Patrasayer, and Ranibandh suffering the highest declines. These findings demonstrate that fragmentation effects extend beyond biodiversity impacts to fundamentally alter the capacity of forests to sustain human populations through ecosystem service provision.

West Bengal's forests, spanning from Sundarbans mangroves through tropical moist deciduous sal forests to subtropical and temperate Himalayan woodlands, harbor extraordinary biodiversity including endemic orchids, medicinal plants, and fauna ranging from Royal Bengal tigers to Asian elephants. This biological richness faces mounting pressures from habitat loss, climate change, and anthropogenic disturbances. The observed decline of nearly 50% in mammalian diversity in patches below

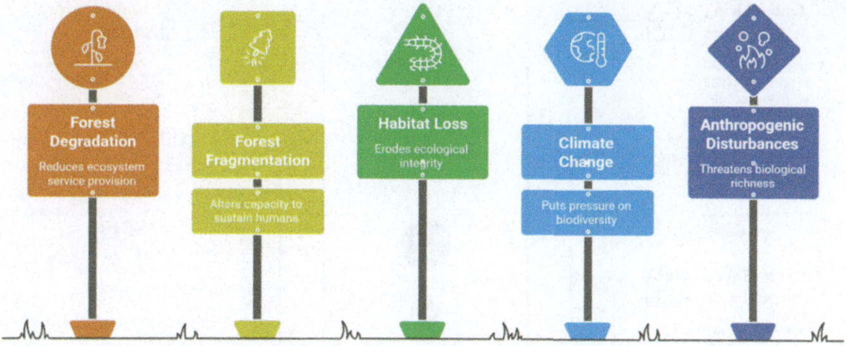

Fig. 10.1 Symmetrical diagram of Forest Ecosystem Services Decline

50 ha, and the loss of over 60% of habitat range for large mammals, signals an erosion of ecological integrity with cascading consequences for ecosystem functionality.

10.6 Livelihood Dependencies and Socio-Economic Dimensions

Chapter 7's detailed examination of forest-livelihood relationships in Binpur-II block reveals the profound dependence of rural communities on forest resources, with non-timber forest products accounting for 47% of total household income. This finding confirms forests as primary economic lifelines for millions of people in West Bengal, particularly members of Scheduled Tribes and Other Backward Classes. The range of utilized forest products 18 species including sal leaves, kendu leaves, mahua flowers, honey, and medicinal plants demonstrates both the diversity of forest contributions and the centrality of specific commodities to rural economies. However, the spatial analysis exposes a troubling paradox: forest proximity does not ensure prosperity. Villages closest to forests, despite higher forest income in absolute terms, exhibit lower overall living standards, with 20% of households in the low-living category compared to 0% in villages farther from forests. This counterintuitive pattern reflects the poverty-environment trap, where dependence on low-value non-timber forest products, poor market access, and limited livelihood diversification combine to perpetuate poverty. In contrast, remote villages with diversified agriculture achieve higher resilience and improved welfare outcomes.

The gendered and caste-based dimensions of forest dependence add further complexity to livelihood patterns. Women dominate small-scale non-timber forest product collection while men control larger commercial activities, with female-headed households showing higher forest dependency due to limited alternatives. Educational and infrastructural deficits compound these poverty cycles, particularly in forest-adjacent villages. The Livelihood Dependency Index demonstrates steep

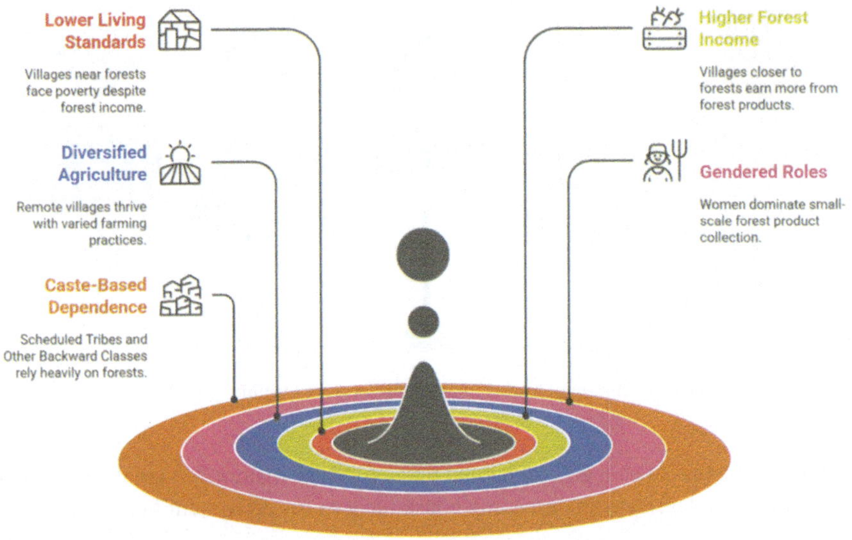

Fig. 10.2 Symmetrical diagram of Forest Dependence and Poverty Paradox

spatial gradients, declining from 94 in interim villages to 27 in outer villages for sal leaves, indicating an adaptive shift from forest-based to agrarian livelihoods with increasing distance from forests.

These findings challenge simplistic assumptions about forest conservation and poverty alleviation (Fig. 10.2). Sustainable solutions must transcend mere protection of forest cover to address the structural inequalities that trap forest-dependent populations in poverty. Value addition through processing, strengthening market linkages, promoting women's self-help groups, and investing in infrastructure and education emerge as critical interventions. Only through such spatially differentiated and socially inclusive strategies can forest conservation align with human dignity and welfare.

10.7 Participatory Forest Management: Achievements and Limitations

The assessment of Joint Forest Management presented in Chap. 8 offers crucial insights into the possibilities and limitations of participatory forest governance. The Arabari case study, representing India's pioneering experiment in collaborative forestry, demonstrates that community participation can achieve remarkable ecological recovery forest crown density increased from below 20% to above 60% within a decade, accompanied by biodiversity recovery and soil stabilization. The subsequent expansion across West Bengal generated tangible benefits: improved canopy cover, enhanced non-timber forest product availability, and livelihood improvements through wage employment and revenue sharing.

However, the quantitative analysis reveals persistent social equity challenges. Only 3% of respondents acknowledged gender equality in Joint Forest Management, with female participation averaging below 2% in decision-making roles. Scheduled Castes and Tribes constituted less than 40% of membership, while dominant castes captured most leadership positions. Satisfaction levels varied dramatically 80% expressed satisfaction with forest restoration outcomes but only 46% felt satisfied with empowerment and usufruct rights. These patterns expose the gap between Joint Forest Management's participatory rhetoric and its practice, where power asymmetries between State Forest Departments and communities persist (Fig. 10.3).

The limitations stem from multiple sources: continued bureaucratic dominance that converts Forest Protection Committees into implementers of department directives rather than autonomous decision-makers; elite capture within communities that concentrates benefits among already privileged groups; inadequate tenure security that undermines long-term community investment; and insufficient integration with the Forest Rights Act of 2006, creating legal ambiguities. The erosion of traditional ecological knowledge through standardized management prescriptions further diminishes the potential for genuine co-management.

Nevertheless, the ecosystem service valuation reveals significant gains an annual increase of 1.15 million US dollars between 1990 and 2023, primarily from expansion of dense forest and enhanced climate regulation and erosion control functions. This ecological success demonstrates that with appropriate institutional arrangements and

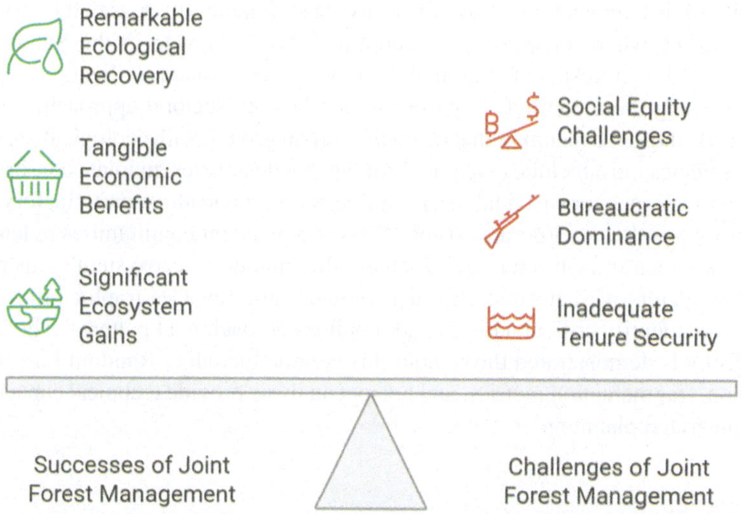

Fig. 10.3 Symmetrical diagram of Balancing Ecological Success with Social Equity in Forest Management

genuine power-sharing, participatory forest management can deliver conservation outcomes while supporting rural livelihoods. The challenge lies in reforming governance structures to ensure equitable participation, transparent benefit distribution, and recognition of community custodianship rights.

10.8 Urban-Forest Conflicts and Integrated Landscape Planning

The analysis of urban expansion presented in Chap. 5 highlights an emerging frontier of forest conservation challenges. As urban populations grow toward 68% globally by 2050, the encroachment of cities upon forests intensifies. In West Bengal, metropolitan expansion around the Kolkata Metropolitan Area has driven approximately 35% forest loss since 1990, while industrial corridors have reduced state forest cover to just 11.8% well below India's national average. The Siliguri case exemplifies these dynamics, with 253% expansion of built-up area occurring largely at the expense of vegetation cover, which declined by 56.29 km^2.

These urban-forest conflicts generate multiple adverse consequences. Ecologically, fragmentation disrupts wildlife corridors and intensifies edge effects, as evidenced by over 150 annual human-elephant conflict incidents around Siliguri. Socially, urban expansion displaces forest-dependent communities, particularly tribal populations, often with minimal compensation a process of ecological marginalization. Economically, failure to account for ecosystem service values estimated at 150 trillion US dollars globally creates perverse incentives favoring short-term development profits over ecological sustainability. The vulnerability revealed by Cyclone Aila in 2009, where mangrove loss amplified storm damage in the Sundarbans, underscores how forest degradation undermines urban climate resilience.

Addressing these conflicts requires moving beyond sectoral approaches toward integrated landscape planning that reconciles urban growth with ecological preservation. Strategies must include ecological zoning that designates inviolate conservation cores, buffer zones for sustainable use, and restoration corridors; participatory planning processes that incorporate voices of forest-dependent communities in land-use decisions; green infrastructure development that maintains ecosystem connectivity within urban matrices; and data-driven governance utilizing Geographic Information Systems to identify critical habitats and monitor encroachment patterns. The spatial analysis tools demonstrated throughout this volume including Random Forest classification, fragmentation metrics, and hotspot analysis provide technical capacity for such integrated planning.

10.9 Climate Change Dimensions and Adaptive Management

Although not detailed in a dedicated chapter, climate change emerges as a cross-cutting theme affecting all dimensions of forest landscapes examined in this volume (Fig. 10.4). Rising temperatures, altered precipitation patterns, and increased frequency of extreme events interact with existing pressures to amplify forest vulnerability. The Sundarbans face particularly acute risks from sea-level rise and cyclone intensification, threatening both mangrove ecosystems and millions of coastal inhabitants. Hill forests in Darjeeling experience shifting altitudinal ranges of species, while the sal-dominated forests of the western plains confront increased moisture stress and fire risk.

Community adaptations to these climate pressures remain inadequately supported by policy and institutional frameworks. Eco-development initiatives described in Chap. 9 attempt to build climate resilience through livelihood diversification promoting apiculture, crab farming, handicrafts, and eco-tourism in the Sundarbans. However, persistent poverty, limited adaptive capacity, and insufficient institutional support constrain effectiveness. The integration of traditional ecological knowledge with scientific climate projections offers potential for developing locally appropriate adaptation strategies, yet systematic mechanisms for such integration remain underdeveloped.

Climate change mitigation and adaptation must be recognized as mutually reinforcing objectives within forest management. Protecting and restoring forests enhances carbon sequestration while maintaining ecosystem services that buffer

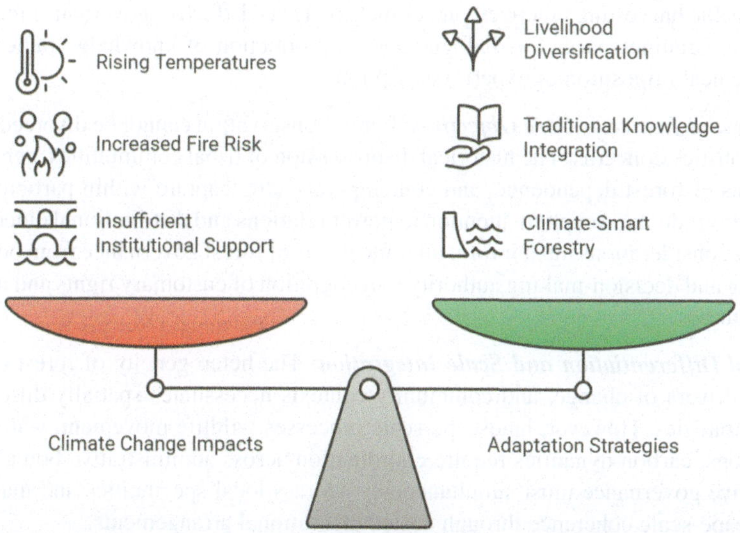

Fig. 10.4 Symmetrical diagram of Balancing Climate Impacts with Adaptation Strategies

communities against climate impacts water regulation, soil conservation, microclimate moderation. The Joint Forest Management finding that community-managed forests achieved 15–25% higher sequestration rates suggests that participatory governance can advance both mitigation and adaptation simultaneously. Future strategies must prioritize climate-smart forestry that combines native species conservation, landscape connectivity, assisted regeneration, and community-based monitoring systems.

10.10 Toward Integrated Social-Ecological Governance

The synthesis of findings across chapters points toward the necessity of integrated social-ecological governance frameworks that transcend conventional divisions between conservation and development, ecology and economy, state and community (Fig. 10.5). Such frameworks must be built upon several foundational principles:

Recognition of Complexity: Forest landscapes constitute complex adaptive systems characterized by non-linear dynamics, threshold effects, and emergent properties. Management approaches must embrace this complexity rather than imposing simplistic models, acknowledging that optimal strategies vary across spatial scales and temporal contexts.

Legitimacy of Multiple Knowledge Systems: Both scientific expertise and traditional ecological knowledge offer valuable insights for forest management. Indigenous communities possess centuries of accumulated understanding about forest ecology, sustainable harvesting practices, and climate patterns. Effective governance requires creating genuine spaces for dialogue and co-production of knowledge rather than hierarchical imposition of expert prescriptions.

Equity and Justice as Core Objectives: Forest conservation cannot be divorced from social justice concerns. The historical dispossession of tribal communities, gendered patterns of forest dependence, and contemporary elite capture within participatory institutions demand explicit attention to power relations and distributional outcomes. Equity considerations must inform all dimensions of forest governance from benefit-sharing and decision-making authority to recognition of customary rights and access to resources.

Spatial Differentiation and Scale Integration: The heterogeneity of forest conditions, drivers of change, and community contexts necessitates spatially differentiated strategies. However, landscape-scale processes wildlife movement, watershed functions, carbon dynamics require coordination across administrative boundaries. Effective governance must simultaneously address local specificities and maintain landscape-scale coherence through nested institutional arrangements.

Fig. 10.5 **a** Local forest-dependent community members engaged in the manual collection of undergrowth and minor forest produce within a densely vegetated woodland, indicating subsistence-based resource extraction practices. **b** Rural women transporting bundled fuelwood along an unpaved village road bordered by plantation forests, reflecting household energy dependence on forest biomass. **c** Open grazing of goats within a forested landscape, illustrating the interaction between livestock rearing practices and forest floor conditions. **d** A traditional mud-walled rural dwelling with basic household activities occurring in the courtyard, highlighting vernacular housing patterns and livelihood conditions in forest-fringe settlements

Adaptive Management and Learning: Given inherent uncertainties in social-ecological systems, management must be conceived as ongoing experimentation and learning rather than implementation of fixed prescriptions. Monitoring systems that track both ecological and social indicators, mechanisms for incorporating feedback into policy revision, and institutional cultures that reward innovation and tolerate calculated risks are essential for adaptive capacity.

Technology Integration with Participatory Processes: Geographic Information Systems, remote sensing, and machine learning offer powerful capabilities for monitoring forest conditions, predicting change trajectories, and identifying priority interventions. However, technology alone cannot substitute for participatory processes. Optimal approaches integrate technological tools with participatory Geographic Information Systems, community-based monitoring, and traditional indicators to create comprehensive assessment frameworks.

Policy Recommendations and Implementation Pathways

Translating these principles into practice requires concrete policy reforms and institutional innovations across multiple domains (Fig. 10.6):

Tenure Security and Rights Recognition: Full implementation of the Forest Rights Act of 2006 is imperative, moving beyond token individual claims to genuine Community Forest Resource governance. As of 2020, only 102 Community Forest Resource titles had been issued across West Bengal a figure that must increase dramatically. Security of tenure provides the foundation for long-term community investment in sustainable management.

Institutional Reform of Joint Forest Management: Joint Forest Management must evolve from its current state-dominated structure toward genuine co-management. This requires legal reforms that recognize Forest Protection Committees as autonomous institutions with legally binding authority rather than advisory bodies; transparent benefit-sharing mechanisms that ensure equitable distribution including timber revenues; mandatory representation quotas for women, Scheduled Castes, and Scheduled Tribes in leadership positions; and conflict resolution mechanisms that provide recourse when disputes arise.

Landscape-Scale Conservation Planning: Protection strategies must shift from isolated protected areas toward landscape-scale approaches that maintain connectivity. Priority actions include mapping and securing critical corridors for elephant

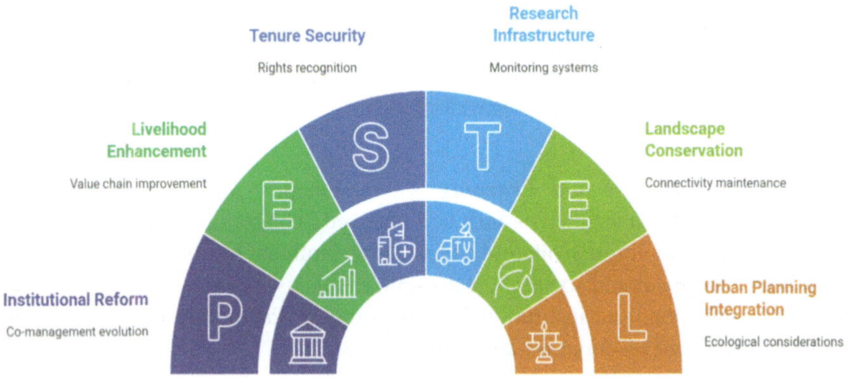

Fig. 10.6 Symmetrical diagram of policy recommendations and implementation pathways

and other wide-ranging species; establishing buffer zones around core conservation areas with regulated sustainable use; restoring degraded forests to enhance landscape connectivity; and coordinating conservation planning across administrative boundaries through watershed or ecoregion frameworks.

Livelihood Enhancement and Value Addition: Breaking the poverty-environment trap requires moving forest-dependent households up value chains through processing infrastructure for non-timber forest products; market linkages that reduce intermediary exploitation; skill development for alternative livelihoods; and microfinance and self-help group support particularly for women. Simultaneously, diversification of rural economies through education, infrastructure investment, and non-farm employment opportunities reduces pressure on forests while improving welfare.

Urban Planning Integration: Managing urban-forest conflicts demands incorporating ecological considerations into urban development planning through mandatory forest impact assessments for infrastructure projects; green belt regulations around urban peripheries; incentive structures for compact urban development rather than sprawl; and compensation mechanisms that capture development gains to fund conservation and community compensation.

Climate-Responsive Forest Management: Adaptation strategies must be mainstreamed across forest policy including species composition shifts toward climate-resilient varieties; fire management protocols appropriate for increasing fire risk; watershed protection to buffer against precipitation variability; and disaster risk reduction that recognizes forest roles in protecting against cyclones, floods, and landslides.

Research and Monitoring Infrastructure: Evidence-based management requires sustained investment in comprehensive forest inventories integrating ground measurements with remote sensing; long-term ecological research sites for understanding forest dynamics; socio-economic monitoring systems tracking livelihood outcomes; and predictive modeling capacity for scenario analysis and decision support.

10.11 Conclusion

West Bengal's Forest landscapes stand at a critical juncture. Centuries of exploitation, fragmentation, and governance failures have degraded what were once continuous, biodiverse ecosystems while simultaneously impoverishing the communities most dependent upon forest resources. Yet the evidence presented throughout this volume also reveals grounds for measured optimism. The Arabari experiment demonstrated that degraded forests can regenerate through community stewardship. Joint Forest Management, despite limitations, has improved forest conditions across significant areas. Geospatial technologies provide unprecedented capacity to monitor change and guide interventions. Most importantly, forest-dependent communities retain both the motivation and knowledge to manage forests sustainably when provided genuine rights and support. The path forward requires fundamental shifts in how forests are conceptualized and governed. Forests must be understood not merely as timber reserves or carbon sinks but as complex social-ecological systems providing multiple services essential for both human welfare and planetary health. Governance must evolve from centralized control toward genuine power-sharing with communities whose lives and livelihoods are intimately connected with forest landscapes. Conservation must be pursued not through exclusion but through inclusive approaches that recognize the rights and dignity of forest-dependent populations.

The ultimate measure of success will not be found solely in hectares of forest cover or tons of carbon sequestered, though these metrics matter. Rather, success must be judged by whether forest landscapes can sustain both biodiversity and thriving human communities, whether indigenous knowledge receives the respect it deserves alongside scientific expertise, whether the most marginalized voices gain genuine influence over decisions affecting their lives, and whether future generations inherit forest ecosystems capable of providing the ecosystem services upon which all life depends. This volume has documented the challenges confronting West Bengal's forests in granular detail. It has also illuminated pathways toward more sustainable and equitable futures. Whether these pathways are followed depends upon political will, institutional reform, resource allocation, and most fundamentally, recognition that forest conservation and human welfare are not competing objectives but deeply interconnected imperatives. The forests of West Bengal, resilient through centuries of transformation, await the governance they deserve governance that honors both ecological integrity and social justice, that integrates traditional wisdom with modern science, and that ensures these magnificent landscapes endure for generations yet to come.

Index

P. K. Shit and M. Hasanuzzaman, *Forest Landscape Dynamics in West Bengal*, Advances in Asian Human-Environmental Research, https://doi.org/10.1007/978-3-032-23700-2

The manufacturer's authorised representative in the EU is Springer Nature Customer Service Centre GmbH, Europaplatz 3, 69115 Heidelberg, Germany. If you have any concerns regarding our products, please contact ProductSafety@springernature.com

Printed and bound by CPI Group (UK) Ltd, Croydon, CR0 4YY
05/10/2026
02228057-0005